# Organic Production and Use of Alternative Crops

*Soil Biochemistry, Volume 1,* edited by A. D. McLaren and G. H. Peterson

*Soil Biochemistry, Volume 2,* edited by A. D. McLaren and J. Skujins

*Soil Biochemistry, Volume 3,* edited by E. A. Paul and A. D. McLaren

*Soil Biochemistry, Volume 4,* edited by E. A. Paul and A. D. McLaren

*Soil Biochemistry, Volume 5,* edited by E. A. Paul and J. N. Ladd

*Soil Biochemistry, Volume 6,* edited by Jean-Marc Bollag and G. Stotzky

*Soil Biochemistry, Volume 7,* edited by G. Stotzky and Jean-Marc Bollag

*Soil Biochemistry, Volume 8,* edited by Jean-Marc Bollag and G. Stotzky

*Soil Biochemistry, Volume 9,* edited by G. Stotzky and Jean-Marc Bollag

*Handbook of Phytoalexin Metabolism and Action*, edited by M. Daniel and R. P. Purkayastha

*Soil–Water Interactions: Mechanisms and Applications, Second Edition, Revised and Expanded*, Shingo Iwata, Toshio Tabuchi, and Benno P. Warkentin

*Stored-Grain Ecosystems*, edited by Digvir S. Jayas, Noel D. G. White, and William E. Muir

*Agrochemicals from Natural Products*, edited by C. R. A. Godfrey

*Seed Development and Germination*, edited by Jaime Kigel and Gad Galili

*Nitrogen Fertilization in the Environment*, edited by Peter Edward Bacon

*Phytohormones in Soils: Microbial Production and Function*, William T. Frankenberger, Jr., and Muhammad Arshad

*Handbook of Weed Management Systems*, edited by Albert E. Smith

*Soil Sampling, Preparation, and Analysis*, Kim H. Tan

*Soil Erosion, Conservation, and Rehabilitation*, edited by Menachem Agassi

*Plant Roots: The Hidden Half, Second Edition, Revised and Expanded*, edited by Yoav Waisel, Amram Eshel, and Uzi Kafkafi

*Photoassimilate Distribution in Plants and Crops: Source–Sink Relationships*, edited by Eli Zamski and Arthur A. Schaffer

*Mass Spectrometry of Soils*, edited by Thomas W. Boutton and Shinichi Yamasaki

*Handbook of Photosynthesis*, edited by Mohammad Pessarakli

*Chemical and Isotopic Groundwater Hydrology: The Applied Approach, Second Edition, Revised and Expanded*, Emanuel Mazor

*Fauna in Soil Ecosystems: Recycling Processes, Nutrient Fluxes, and Agricultural Production*, edited by Gero Benckiser

*Soil and Plant Analysis in Sustainable Agriculture and Environment*, edited by Teresa Hood and J. Benton Jones, Jr.

*Seeds Handbook: Biology, Production, Processing, and Storage*, B. B. Desai, P. M. Kotecha, and D. K. Salunkhe

*Modern Soil Microbiology*, edited by J. D. van Elsas, J. T. Trevors, and E. M. H. Wellington

*Growth and Mineral Nutrition of Field Crops: Second Edition*, N. K. Fageria, V. C. Baligar, and Charles Allan Jones

*Fungal Pathogenesis in Plants and Crops: Molecular Biology and Host Defense Mechanisms*, P. Vidhyasekaran

*Plant Pathogen Detection and Disease Diagnosis*, P. Narayanasamy

*Agricultural Systems Modeling and Simulation*, edited by
Robert M. Peart and R. Bruce Curry

*Agricultural Biotechnology*, edited by Arie Altman

*Plant–Microbe Interactions and Biological Control*, edited by
Greg J. Boland and L. David Kuykendall

*Handbook of Soil Conditioners: Substances That Enhance
the Physical Properties of Soil*, edited by Arthur Wallace
and Richard E. Terry

*Environmental Chemistry of Selenium*, edited by
William T. Frankenberger, Jr., and Richard A. Engberg

*Principles of Soil Chemistry: Third Edition, Revised and Expanded*,
Kim H. Tan

*Sulfur in the Environment*, edited by Douglas G. Maynard

*Soil–Machine Interactions: A Finite Element Perspective*, edited by
Jie Shen and Radhey Lal Kushwaha

*Mycotoxins in Agriculture and Food Safety*, edited by Kaushal K.
Sinha and Deepak Bhatnagar

*Plant Amino Acids: Biochemistry and Biotechnology*, edited by
Bijay K. Singh

*Handbook of Functional Plant Ecology*, edited by Francisco I.
Pugnaire and Fernando Valladares

*Handbook of Plant and Crop Stress: Second Edition, Revised
and Expanded*, edited by Mohammad Pessarakli

*Plant Responses to Environmental Stresses: From Phytohormones
to Genome Reorganization*, edited by H. R. Lerner

*Handbook of Pest Management*, edited by John R. Ruberson

*Environmental Soil Science: Second Edition, Revised and Expanded*,
Kim H. Tan

*Microbial Endophytes*, edited by Charles W. Bacon
and James F. White, Jr.

*Plant–Environment Interactions: Second Edition*, edited by
Robert E. Wilkinson

*Microbial Pest Control*, Sushil K. Khetan

*Soil and Environmental Analysis: Physical Methods, Second Edition,
Revised and Expanded*, edited by Keith A. Smith
and Chris E. Mullins

*The Rhizosphere: Biochemistry and Organic Substances at the
Soil–Plant Interface*, Roberto Pinton, Zeno Varanini,
and Paolo Nannipieri

*Woody Plants and Woody Plant Management: Ecology, Safety,
and Environmental Impact*, Rodney W. Bovey

*Metals in the Environment*, M. N. V. Prasad

*Plant Pathogen Detection and Disease Diagnosis: Second Edition, Revised and Expanded*, P. Narayanasamy

*Handbook of Plant and Crop Physiology: Second Edition, Revised and Expanded*, edited by Mohammad Pessarakli

*Environmental Chemistry of Arsenic*, edited by William T. Frankenberger, Jr.

*Enzymes in the Environment: Activity, Ecology, and Applications*, edited by Richard G. Burns and Richard P. Dick

*Plant Roots: The Hidden Half, Third Edition, Revised and Expanded*, edited by Yoav Waisel, Amram Eshel, and Uzi Kafkafi

*Handbook of Plant Growth: pH as the Master Variable*, edited by Zdenko Rengel

*Biological Control of Major Crop Plant Diseases* edited by Samuel S. Gnanamanickam

*Pesticides in Agriculture and the Environment*, edited by Willis B. Wheeler

*Mathematical Models of Crop Growth and Yield*, , Allen R. Overman and Richard Scholtz

*Plant Biotechnology and Transgenic Plants*, edited by Kirsi-Marja Oksman Caldentey and Wolfgang Barz

*Handbook of Postharvest Technology: Cereals, Fruits, Vegetables, Tea, and Spices*, edited by Amalendu Chakraverty, Arun S. Mujumdar, G. S. Vijaya Raghavan, and Hosahalli S. Ramaswamy

*Handbook of Soil Acidity*, edited by Zdenko Rengel

*Humic Matter in Soil and the Environment: Principles and Controversies*, edited by Kim H. Tan

*Molecular Host Plant Resistance to Pests*, edited by S. Sadasivam and B. Thayumanayan

*Soil and Environmental Analysis: Modern Instrumental Techniques, Third Edition*, edited by Keith A. Smith and Malcolm S. Cresser

*Chemical and Isotopic Groundwater Hydrology, Third Edition*, edited by Emanuel Mazor

*Agricultural Systems Management: Optimizing Efficiency and Performance*, edited by Robert M. Peart and W. David Shoup

*Physiology and Biotechnology Integration for Plant Breeding*, edited by Henry T. Nguyen and Abraham Blum

*Global Water Dynamics: Shallow and Deep Groundwater: Petroleum Hydrology: Hydrothermal Fluids, and Landscaping*, , edited by Emanuel Mazor

*Principles of Soil Physics*, edited by Rattan Lal

*Seeds Handbook: Biology, Production, Processing, and Storage, Second Edition*, Babasaheb B. Desai

*Field Sampling: Principles and Practices in Environmental Analysis,*
edited by Alfred R. Conklin

*Sustainable Agriculture and the International Rice-Wheat System,*
edited by Rattan Lal, Peter R. Hobbs, Norman Uphoff,
and David O. Hansen

*Plant Toxicology, Fourth Edition,* edited by Bertold Hock
and Erich F. Elstner

*Drought and Water Crises: Science, Technology, and Management
Issues,* edited by Donald A. Wilhite

*Soil Sampling, Preparation, and Analysis, Second Edition,* Kim H. Tan

*Climate Change and Global Food Security,* edited by Rattan Lal,
Norman Uphoff, B. A. Stewart, and David O. Hansen

*Handbook of Photosynthesis, Second Edition,* edited by
Mohammad Pessarakli

*Environmental Soil-Landscape Modeling: Geographic Information
Technologies and Pedometrics,* edited by Sabine Grunwald

*Water Flow In Soils, Second Edition,* Tsuyoshi Miyazaki

*Biological Approaches to Sustainable Soil Systems,* edited by
Norman Uphoff, Andrew S. Ball, Erick Fernandes, Hans Herren,
Olivier Husson, Mark Laing, Cheryl Palm, Jules Pretty, Pedro
Sanchez, Nteranya Sanginga, and Janice Thies

*Plant–Environment Interactions, Third Edition,* edited by Bingru Huang

*Biodiversity In Agricultural Production Systems,* edited by
Gero Benckiser and Sylvia Schnell

*Organic Production and Use of Alternative Crops,* Franc Bavec
and Martina Bavec

# Organic Production and Use of Alternative Crops

## Franc Bavec
## Martina Bavec

**CRC Press**
Taylor & Francis Group
Boca Raton London New York

CRC Press is an imprint of the
Taylor & Francis Group, an **informa** business
A TAYLOR & FRANCIS BOOK

CRC Press
Taylor & Francis Group
6000 Broken Sound Parkway NW, Suite 300
Boca Raton, FL 33487-2742

© 2007 by Taylor & Francis Group, LLC
CRC Press is an imprint of Taylor & Francis Group, an Informa business

First issued in paperback 2019

No claim to original U.S. Government works

ISBN 13: 978-0-367-45353-4 (pbk)
ISBN 13: 978-1-57444-617-3 (hbk)

---

### Library of Congress Cataloging-in-Publication Data

---

Bavec, Franc.
    Organic production and use of alternative crops / Franc Bavec, Martina Bavec.
      p. cm. -- (Books in soils, plants, and the environment ; 116)
    Includes bibliographical references and index.
    ISBN-13: 978-1-57444-617-3
    ISBN-10: 1-57444-617-7
    1. Organic farming. 2. Alternative agriculture. 3. Food crops. I. Bavec, Martina. II. Title. III. Series.

S605.5.B397 2006
631.5'84--dc22
                                                    2006006060

**Visit the Taylor & Francis Web site at**
**http://www.taylorandfrancis.com**

**and the CRC Press Web site at**
**http://www.crcpress.com**

# Preface

Recent developments in organic agriculture have contributed significantly to the increasing sustainability of the countryside, animal welfare, and human awareness. Changes to the organic crop management system respect variegated biodiversity, natural resources at regional production sites, and natural products. Human- and environment-friendly crop production and food processing systems have replaced chemicals (fertilizers, pesticides, pharmaceuticals, growth regulators, and unhealthy additives), gene modified organisms (GMOs), and nonresistant cultivars to diseases. Contrary to "pro et contra" thinking about human health and the consumption of genetically modified food, intensive crop production and new biotechnological discoveries directly support harmful agricultural practices. The need for insect resistance in genetically modified crops is obligatory for intensive crop production, like non-crop rotation systems under monoculture, with numerous negative environmental and biological consequences (e.g., the unbalanced biodiversity of crops without naturally controlled organisms). Intensive production systems need more efficient pesticides because of weeds, pests, and diseases tolerance to active chemical substances; however, these will kill living soil over long-term periods. Additional moral and economical questions are raised by "terminator" genes, the long-term contamination of natural plants by genetically modified genes, and so on.

Intensive agriculture has become the factor that evidently reduces the number of utilized crops, and, as a consequence, this limitation decreases possibilities for further development. The need for variegated biodiversity must be passed on in the new millennium not only for wide natural protection purposes, but also for the agricultural purpose of discovering alternative (old, ancient, neglected, disregarded, or new) crops. Alternative crops are a rich natural resource of many nutritional compounds, which are often limited to products from just a few main crops produced in our fields. The introduction of alternative crops into rotation will contribute to more natural production systems. Alternative crops can also help to reduce natural oversensitivity by sowing more resistant genotypes to plant diseases, increasing the population of natural predators, changing weed population, and helping us to produce food without synthetic pesticides. On this basis, organic crop management represents a very important option for environmentally friendly crop production and special "organic" products. In addition, the food based on alternative crops is rich in essential dietary and health compounds.

The authors of this book have extensive practical experience as organic farmers in growing many alternative crops in temperate climatic conditions. Without this practice, it would be much more difficult to write about the organic production of alternative crops, since there is not much scientific data available on this specific topic. This is the main reason for statements based on our own experiences with the organic growing of alternative plants, though they may not always be well supported by research results. Working as advisers and as university teachers has taught us that, in many cases, obvious statements must be expressed; otherwise, they are underestimated. Since there are many alternative crops covered in this book, with many similar growing techniques, general statements about organic growth are written in the introductory chapters in order to avoid the repetition of basic information. Further information about growing techniques are usually presented in detail for one plant of the group (e.g., spelt wheat compared to other alternative wheat species) or, based on the authors' opinions, for crops with similar perspectives in organic farming system (e.g., oil pumpkins, buckwheat, and so forth).

Some alternative crops in organic farming also appear as main crops in conventional agriculture; these are described here. Personal experience in the processing of alternative crops, along with the management of the processing and marketing of organic products and our own research and close cooperation with many small-scale processing activities on farms and in the food industry, has given us an overview of major problems and a basis for the utilization of alternative crops. Primarily, recently published research about the utilization of alternative crops has been taken into account as an initiative for further research and development of organic products, their nutritional value, the traditional sense for human consumption, and other developing uses.

We hope that all who like nature and respect natural food, and all who work in agriculture, especially in organic crop production and new market niches, will find something of interest in this book. We especially hope that this book will be useful to the new generation of students of agriculture, food processing, nutrition, food science, and marketing, as well as to teachers and advisers in both developed and developing countries.

**Franc Bavec & Martina Bavec**

# *Author/Editor Page*

*Dr. Franc Bavec,* born in 1959, is a professor of field crops and organic agriculture at the University of Maribor, Slovenia. He has authored and coauthored 19 research papers, 3 books, and many professional papers. His knowledge about organic agriculture based on national and international research projects includes sustainable agricultural systems, organic farming, production, utilization of alternative crops, and experiences from his own organic farm. He has coordinated international intensive study programs' "Alternatives for Organic Production of Field Crops" and has been invited to lecture on the topic of alternative crops. He is also active as a national representative to the European Society for Agronomy and as vice-rector of the university where he teaches.

*Dr. Martina Bavec,* born in 1962, is assistant professor of vegetable and field crops and organic agriculture at the University of Maribor, Slovenia. She has been an adviser for vegetable production and organic farming, as well as the main supporter for establishing a national inspection and certification body for organic agriculture. She has led many national and international projects, edited the main national book about organic agriculture, and authored and coauthored 11 research papers and numerous professional papers. She is a very active presenter of organic agriculture in the community and a key representative of the BSc, MSc, and PhD study programs of organic agriculture at the university where she teaches.

# Contents

**Chapter 1   Introduction** ...................................................................... 1
1.1   What is organic agriculture? ............................................................ 1
    1.1.1   Definition ............................................................................... 2
    1.1.2   Development of organic farming ........................................ 5
    1.1.3   Standards ............................................................................... 8
    1.1.4   Some data about organic agriculture ................................ 9
1.2   Organic food ...................................................................................... 12
    1.2.1   Organic food and legislation ............................................. 13
    1.2.2   Quality of organic food ...................................................... 15
1.3   Multifunctionality of organic agriculture .................................... 17
    1.3.1   Sociological aspects ............................................................. 17
    1.3.2   Biodiversity .......................................................................... 18
    1.3.3   Environmental impacts ....................................................... 20
1.4   Organic crop production .................................................................. 22
    1.4.1   Rules in organic crop production ...................................... 22
    1.4.2   Organic crop management ................................................. 25
        1.4.2.1   Influence of precrop ............................................. 26
        1.4.2.2   Intercropping ......................................................... 27
1.5   Contribution of organic agriculture to the conservation of
alternative crops and their utilization development .................... 28
References ..................................................................................................... 33

**Chapter 2   Cereals** ................................................................................ 37
2.1   Spelt .................................................................................................... 37
    2.1.1   Introduction ......................................................................... 37
    2.1.2   Botany ................................................................................... 38
        2.1.2.1   Systematic and genotypes ................................... 38
        2.1.2.2   Morphology ........................................................... 39
    2.1.3   Growth and ecology ........................................................... 39
    2.1.4   Organic cultivation practice .............................................. 39
    2.1.5   Harvesting ............................................................................ 42
    2.1.6   Chemical composition, nutritional, and health value ..... 42
    2.1.7   Processing ............................................................................. 45
2.2   Einkorn ............................................................................................... 47

|         | 2.2.1 | Introduction | 47 |
|         | 2.2.2 | Botany | 49 |
|         | 2.2.3 | Cultivation practices | 49 |
|         | 2.2.4 | Chemical composition and processing | 49 |
| 2.3 | Emmer | | 50 |
|         | 2.3.1 | Origin, botany, and history | 50 |
|         | 2.3.2 | Production | 50 |
|         | 2.3.3 | Utilization | 51 |
| 2.4 | Kamut | | 51 |
| 2.5 | Triticale | | 52 |
|         | 2.5.1 | Introduction | 52 |
|         | 2.5.2 | History, taxonomy, and cultivars | 53 |
|         | 2.5.3 | Morphology | 54 |
|         | 2.5.4 | Production and yielding | 54 |
|         | 2.5.5 | Growth and ecology | 54 |
|         | 2.5.6 | Organic cultivation practice | 55 |
|         |        | 2.5.6.1 Crop rotation | 55 |
|         |        | 2.5.6.2 Soil | 56 |
|         |        | 2.5.6.3 Sowing and plant cultivation practice | 56 |
|         | 2.5.7 | Nutritional and health value | 57 |
|         | 2.5.8 | Processing | 57 |
| 2.6 | Intermediate wheatgrass | | 58 |
| References | | | 59 |

| Chapter 3 | Pseudocereals (without millets) | | 65 |
| 3.1 | Buckwheat | | 65 |
|         | 3.1.1 | Introduction | 65 |
|         | 3.1.2 | Botany | 66 |
|         |        | 3.1.2.1 Taxonomy | 66 |
|         |        | 3.1.2.2 Morphology | 66 |
|         | 3.1.3 | Production and yielding | 68 |
|         | 3.1.4 | Growth and ecology | 68 |
|         | 3.1.5 | Organic cultivation practice | 70 |
|         |        | 3.1.5.1 Crop rotation | 70 |
|         |        | 3.1.5.2 Soil, plowing, and presowing preparations | 70 |
|         |        | 3.1.5.3 Sowing and crop cultivation | 71 |
|         |        | 3.1.5.4 Fertilization | 72 |
|         | 3.1.6 | Harvesting, handling, and storage | 72 |
|         | 3.1.7 | Nutritional and health value | 73 |
|         |        | 3.1.7.1 Nutritional value | 73 |
|         |        | 3.1.7.2 Health value | 74 |
|         |        | 3.1.7.3 Processing | 76 |
|         |        | 3.1.7.4 Milling | 76 |
| 3.2 | Quinoa | | 78 |

|       | 3.2.1 | Introduction | 78 |
|       | 3.2.2 | Botany | 79 |
|       |       | 3.2.2.1 Taxonomy | 79 |
|       |       | 3.2.2.2 Morphology | 79 |
|       | 3.2.3 | Growth and Ecology | 81 |
|       | 3.2.4 | Organic cultural practice | 83 |
|       |       | 3.2.4.1 Crop rotation, pests, and diseases | 83 |
|       |       | 3.2.4.2 Sowing and intercultural operations | 83 |
|       | 3.2.5 | Harvesting | 84 |
|       | 3.2.6 | Chemical composition, nutritional and health value | 84 |
|       | 3.2.7 | Gastronomy and suggestions for homemade food | 87 |
|       | 3.2.8 | Processing | 87 |
| 3.3   | Grain amaranths | | 88 |
|       | 3.3.1 | Introduction | 88 |
|       | 3.3.2 | Botany of grain amaranths | 89 |
|       | 3.3.3 | Ecology | 90 |
|       | 3.3.4 | Disposition for organic cultural practice | 92 |
|       |       | 3.3.4.1 Growth patterns | 92 |
|       |       | 3.3.4.2 Cultivation systems | 92 |
|       | 3.3.5 | Harvesting and yielding | 94 |
|       | 3.3.6 | Nutritional value | 94 |
|       | 3.3.7 | Food processing | 95 |
|       | 3.3.8 | Health value | 98 |
| 3.4.  | Wild rice | | 98 |
|       | 3.4.1 | Introduction | 98 |
|       | 3.4.2 | Biology | 98 |
|       | 3.4.3 | Growth and cultivation | 99 |
|       | 3.4.4 | Nutritional value and utilization | 99 |
| References | | | 100 |

| **Chapter 4** | **Millets** | | **109** |
|---|---|---|---|
| 4.1   | Proso millet | | 110 |
|       | 4.1.1 | Introduction | 110 |
|       | 4.1.2 | Botany | 111 |
|       |       | 4.1.2.1 Morphology | 111 |
|       | 4.1.3 | Climate conditions | 111 |
|       | 4.1.4 | Cultural practice | 112 |
|       | 4.1.5 | Utilization | 114 |
| 4.2   | Foxtail, foxtail millet | | 114 |
|       | 4.2.1 | Introduction | 114 |
|       | 4.2.2 | Biological characteristics | 114 |
|       | 4.2.3 | Cultivars | 115 |
|       | 4.2.4 | Cultural practice | 116 |
|       | 4.2.5 | Utilization | 116 |

4.3 Pearl millet.........................................................................................117
    4.3.1 Introduction ...........................................................................117
    4.3.2 Growth conditions and cultural practice.............................117
    4.3.3 Nutritional value....................................................................118
    4.3.4 Pearl millet processing and utilization..............................119
4.4 Finger millet .......................................................................................119
    4.4.1 Introduction ...........................................................................119
    4.4.2 Biology....................................................................................120
    4.4.3 Growth and production characteristics...............................120
    4.4.4 Nutritional value and utilization ........................................120
4.5 White fonio ..........................................................................................121
    4.5.1 Introduction ...........................................................................121
    4.5.2 Biology....................................................................................121
    4.5.3 Cultivation and utilization ..................................................122
4.6 Barnyard millet ..................................................................................123
4.7 Koda millet ..........................................................................................123
4.8 Little millet...........................................................................................124
References .....................................................................................................124

Chapter 5   Alternative oil plants.............................................................127
5.1 Oil (seed) pumpkins...........................................................................127
    5.1.1 Introduction .........................................................................127
    5.1.2 Botany and growth stages...................................................127
    5.1.3 Production..............................................................................129
    5.1.4 Soil and climate....................................................................131
    5.1.5 Cultural practice....................................................................131
    5.1.6 Harvesting, handling, and storage......................................134
    5.1.7 Nutritional value...................................................................134
    5.1.8 Gastronomy and health value .............................................135
    5.1.9 Processing..............................................................................136
5.2 Camelina ..............................................................................................136
    5.2.1 Introduction .........................................................................136
    5.2.2 Utilization, nutritional and health value ...........................136
    5.2.3 Botany and genotypes..........................................................137
    5.2.4 Morphology ...........................................................................137
    5.2.5 Production..............................................................................138
        5.2.5.1 Utilization .................................................................139
        5.2.5.2 Storage......................................................................139
5.3 Safflower...............................................................................................139
    5.3.1 Introduction ..........................................................................139
    5.3.2 Botany ....................................................................................140
    5.3.3 Ecology...................................................................................141
    5.3.4 Possibilities for organic production .....................................142
    5.3.5 Utilization...............................................................................145

5.4 White mustard...................................................................................146
    5.4.1 Introduction .............................................................................146
    5.4.2 Utilization..................................................................................146
    5.4.3 Botany and genotypes.............................................................147
    5.4.4 Growth.......................................................................................147
    5.4.5 Production.................................................................................148
    5.4.6 Processing of new organic products.....................................149
    5.4.7 Production for green salad......................................................150
5.5 (Garden) poppy..............................................................................150
    5.5.1 Introduction .............................................................................150
    5.5.2 Morphology ..............................................................................151
    5.5.3 Growth stages...........................................................................153
    5.5.4 Production and cultivars ........................................................153
    5.5.5 Growth characteristics and organic cultural practice...........154
        5.5.5.1 Oil (seed) poppy production techniques..................154
    5.5.6 Harvesting and yield...............................................................156
    5.5.7 Use, nutritional composition, and research results
           about opiates in the seeds .................................................156
References .................................................................................................158


Chapter 6  Alternative fiber, root, and tuber crops ...........................163
6.1 Industrial and edible-seed hemp .................................................163
    6.1.1 Introduction .............................................................................163
        6.1.1.1 History.........................................................................163
    6.1.2 Botany and ecotypes................................................................166
    6.1.3 Plant morphology and anatomy of stalk..............................166
        6.1.3.1 Stem.............................................................................167
    6.1.4 Ecology of growth and development.....................................170
    6.1.5 Cultivation practice .................................................................172
        6.1.5.1 Cultivars......................................................................172
    6.1.6 Harvesting, storage, and processing......................................174
    6.1.7 Utilization: special organic products.....................................175
        6.1.7.1 Hemp seed..................................................................177
6.2 Flax...................................................................................................178
    6.2.1 Introduction .............................................................................178
    6.2.2 Botany and cultivars................................................................179
    6.2.3 Morphology ..............................................................................179
    6.2.4 Growth and development stages ...........................................180
        6.2.4.1 Growth conditions.....................................................181
        6.2.4.2 Cultivation practice....................................................182
        6.2.4.3 Sowing and cultivation practice ..............................183
    6.2.5 Harvesting................................................................................184
    6.2.6 Utilization..................................................................................184
6.3 Jerusalem artichoke .......................................................................186

| | 6.3.1 | Introduction and crop description | 186 |
| | 6.3.2 | Growth requirements and organic cultivation | 187 |
| | 6.3.3 | The perspective of the use of Jerusalem artichoke also produced as organic product | 188 |
| 6.4 | Sweet potato | | 189 |
| | 6.4.1 | Introduction | 189 |
| | 6.4.2 | Botany | 190 |
| | | 6.4.2.1 Climatic characteristics, growth, and development | 191 |
| | 6.4.3 | Cultivation practice | 191 |
| | 6.4.4 | Plant reproduction | 193 |
| | 6.4.5 | Nutritional value and food processing | 193 |
| | 6.4.6 | Health value | 194 |
| References | | | 195 |

| Chapter 7 | **Legumes** | | **201** |
| 7.1 | Chickpea | | 202 |
| | 7.1.1 | Introduction | 202 |
| | 7.1.2 | Botany | 203 |
| | 7.1.3 | Climatic conditions, growth, and development | 203 |
| | 7.1.4 | Cultivation practice | 205 |
| | 7.1.5 | Some remarks on the nutritional value of chickpea and other grain legumes | 207 |
| 7.2 | Groundnut | | 207 |
| | 7.2.1 | Introduction | 207 |
| | 7.2.2 | Climate, growth, and development | 208 |
| | 7.2.3 | Morphology | 208 |
| | 7.2.4 | Genotypes | 208 |
| | 7.2.5 | Cultivation practice | 209 |
| | 7.2.6 | Nutritional value and use | 210 |
| | 7.2.7 | Dangerous aflatoxins | 211 |
| 7.3 | Soybean: small attention to the important crop | | 211 |
| | 7.3.1 | Introduction into utilization | 211 |
| | 7.3.2 | Genotypes | 213 |
| | 7.3.3 | Morphology | 213 |
| | 7.3.4 | Susceptibility to climate circumstances | 215 |
| | 7.3.5 | Cultivation according to organic guidelines | 215 |
| | | 7.3.5.1 Fertilization | 216 |
| | 7.3.6 | Harvesting | 217 |
| 7.4 | Vigna: a few words about a widely spread genus | | 217 |
| References | | | 221 |

**Chapter 8   Traditional and new kinds of food from alternative crops** ...................................................................... **225**

8.1   Homemade food from buckwheat.............................................. 225

    8.1.1   Buckwheat bread.................................................................. 225

        8.1.1.1   Standard recipe .................................................... 225

    8.1.2   Dishes made from buckwheat groats ................................ 226

        8.1.2.1   Buckwheat kasha................................................. 226

        8.1.2.2   Buckwheat dumplings......................................... 227

        8.1.2.3   Baked buckwheat kasha...................................... 227

        8.1.2.4   Salad from buckwheat groats............................. 227

        8.1.2.5   Chicken, duck, or turkey filled with buckwheat groats ................................................. 227

        8.1.2.6   Soups ..................................................................... 227

        8.1.2.7   Noodles ................................................................. 228

        8.1.2.8   Homemade noodles ............................................. 228

        8.1.2.9   Slovenian ûganci, Italian polenta, and Japanese soba-gaki methods.............................. 228

        8.1.2.10   Cooked buckwheat rolls (with cottage cheese or walnuts)................................................ 228

        8.1.2.11   Buckwheat pancakes........................................... 229

        8.1.2.12   Buckwheat cake .................................................. 229

8.2   Homemade food from amaranths ............................................. 229

    8.2.1   Popped seeds........................................................................ 229

    8.2.2   Bread .................................................................................... 230

    8.2.3   Pasta and noodle products................................................ 230

8.3   Homemade food from millet...................................................... 230

    8.3.1   Millet kasha with ground meat ........................................ 230

References ............................................................................................... 230

Index......................................................................................................... 231

# chapter one

# Introduction

## 1.1 What is organic agriculture?

Organic farming has become the most highly valued method of sustainable production in the agriculture and food trade. The time when organic farming was a viable option only for people with alternative views on environmental preservation, wildlife protection, and health-food production is long gone. Today if a farmer is looking for long-term farming solutions and the production of safe, high-quality food, organic farming is the best option.

Due to consumer fear—caused by the increasing potential for agricultural products to carry diseases, such as "mad cow," or contain harmful additives, such as pesticides, hormones, nitrates, and other unnatural ingredients—and doubts about technological improvements, such as genetically modified or radiated food, there is an ever-increasing demand for healthy, high-quality food. Consumers also want more information about production methods than they have in the past and they are willing to pay more for these improvements.

Increased consumer awareness of food-safety issues and environmental concerns has contributed to the growth in organic farming over the last few years. Although there is still a small percentage of Utilized Agricultural Area (UAA) in the European Union (EU), organic farming represented around 3% of the EU's total UAA in 2000 and in the United States, farmers in 48 states dedicated 2.3 million acres of cropland and pastures to organic production systems in 2001. Organic farming has truly developed into one of the most dynamic agricultural sectors. In the EU, the organic-farming sector grew by about 25% a year between 1993 and 1998 and, since 1998, is estimated to have grown by around 30% a year, although in some member states it seems to have reached a plateau [1, 2].

The growing organic market is now an important factor in farmers' decisions to convert from conventional farming. Additional public awareness of the harmful impact of intensive agricultural production—soil pollution from pesticides and heavy metals, underground water pollution from nitrates and pesticides, decreased biodiversity, and vanishing plant and animal species—demands more responsibility from farmers concerning their attitudes toward nature. Inclusion of rare and alternative field crops has added value to the diversification of crop rotation and preservation of biodiversity. Compared with other agricultural practices organic farming is clearly the best choice; it provides safe, high-quality food and leaves a smaller mark on the environment than most conventional methods. These differences will be further explored in this introductory chapter.

## 1.1.1  Definition

Organic agriculture has become an important alternative to intensive, conventional farming. International understanding of organic agriculture as a production and processing management system is based on the standards set by the International Federation of Organic Agriculture Movements (IFOAM), national and community laws and programs, and production guidelines for farmers and their associations supported by society and public policy. The basic principles of organic agriculture are to protect nature, prevent harmful influences on the environment and the landscape, develop biological diversity, care for drinking water, and produce high-quality food. For these reasons, the basic principles of organic agriculture around the world, according to IFOAM, are preserving soil fertility, using natural mechanisms in ecosystems, improving natural resources, minimizing environmental pollution, maintaining animal welfare, and connecting production to farmland.

The organic-farming system does not allow the use of synthetic chemical insecticides, fungicides, herbicides, and fertilizers; synthetic additives and growth regulators; antibiotics and hormones for increased animal production; or gene manipulation of any organism that is a part of food production or processing.

The concept of organic farming can be defined in many ways. One of them is the definition developed by the Codex Alimentarius based on contributions from experts from all over the world. According to the Codex, organic farming involves holistic production-management systems for crops and livestock which emphasize management practices (bought outside the farm) inputs. This is accomplished by using cultural, biological, and mechanical methods instead of synthetic materials whenever possible [3*]. The Codex guidelines specify that an organic production system is designed to:

- Enhance biological diversity within the system
- Increase soil biological activity
- Maintain long-term soil fertility
- Recycle plant and animal waste to return nutrients to the land, thus minimizing the use of nonrenewable resources
- Rely on renewable resources in locally organized agricultural systems

---

* *Organic Farming*, Office for Official Publications of the European Commission, © European Communities, 2001.

- Promote the healthy use of soil, water, and air, as well as minimize all forms of pollution that may result from agricultural practices
- Handle agricultural products with an emphasis on careful processing methods to maintain the organic integrity and vital qualities of the product at all stages
- Establish itself on any existing farm through a conversion period, the appropriate length of which is determined by site-specific factors, such as the history of the land and the type of crops and livestock to be produced [4]

The organic-agriculture sector is the fastest growing food sector today. The reasons for this include the withdrawal of government subsidies for agricultural inputs, the introduction of policies that are favorable toward organic agriculture, the controversial food and environmental safety debate on genetic modification, and the crisis provoked by dioxin-contaminated food, commonly known as mad-cow and foot-and-mouth disease [5].

The reasons for organic farming expansion are numerous and they vary from country to country. Consumer interest has grown in response to repeated food-safety scares and animal-welfare concerns, among others regarding the impact of industrial agriculture on the environment. Producers have also been attracted because of environmental concerns as well as the potential health impacts of using agrochemicals and the economics of organic production relative to conventional agriculture. This latter factor has been affected by the fact that many governments are encouraging more producers to adopt organic techniques [6*].

With the continuous growth of the organic sector, and the challenges and opportunities that come with that growth, the 2005 IFOAM General Assembly in Adelaide came to the conclusion that the basic values—the fundamental underpinnings of organic agriculture—needed further reflection and discussion. The approved "Principles of Organic Agriculture" consist of four objectives upon which organic agriculture is based:

- The Principle of Health—Organic Agriculture should sustain and enhance the health of soil, plant, animal, and human as one and indivisible.
- The Principle of Ecology—Organic Agriculture should be based on living ecological systems and cycles, work with them, emulate them, and help sustain them.
- The Principle of Fairness—Organic Agriculture should build on relationships that ensure fairness with regard to the common environment and life opportunities.

- The Principle of Care—Organic Agriculture should be managed in a precautionary and responsible manner to protect the health and well being of current and future generations and the environment.

Principles apply to agriculture in the broadest sense, including the way people tend soil, water, plants, and animals in order to produce, prepare, and distribute goods. They concern the way people interact with living landscapes, relate to one another, and shape the legacy of future generations [7].

Organic farming is a promising agricultural method with positive effects on the human ecological and social environment [8*]. Contemporary organic farming is based on a number of approaches, which have blended over time to produce the current school of thought. A modern definition of organic farming states that the aim is: "to create integrated, humane, environmentally and economically sustainable production systems, which maximize reliance on farm-derived renewable resources and the management of ecological and biological processes and interactions, so as to provide acceptable levels of crop, livestock and human nutrition, protection from pests and disease, and an appropriate return to the human and other resources" [9].

Organic farming differs from other farming systems in a number of ways. It favors renewable resources and recycling—the return of nutrients found in waste products back to the soil. Where livestock is concerned, meat and poultry production is regulated with particular concern for animal welfare and the use of natural foodstuffs. Organic farming respects the environment's own systems for controlling pests and disease in raising crops and livestock and avoids the use of synthetic pesticides, herbicides, chemical fertilizers, growth hormones, antibiotics, or gene manipulation. Instead, organic farmers use a range of techniques that help sustain ecosystems and reduce pollution [1].

The goals of organic farming convey a predominantly farm-oriented and production-practice perspective. However, the implications reach further than the farm. Practices relevant to consumers do not include synthetic pesticides and industrial fertilizers, genetically modified organisms (GMOs) in food production, synthetic growth enhancers, or color additives in fodder. These same practices include only medicinal use of antibiotics, security zones to separate organic farmland from conventional farmland and roads, and restricted use of additives in processed foods. Practices relevant to environmental concerns include maximal use of local resources and recycling of organic material, avoiding nitrogen leaching, and reducing the use of fossil resources in transportation. Ethical considerations include enhancing biological diversity, keeping high ethical standards in animal husbandry, acknowledging the animals' innate nature and needs, and working for a socially just food-system. Thus, these considerations also reflect a wider social and

* Reprinted from Journal of Sustainable Agriculture, 26 (1), Vogl, C.R., Kilcher, L., and Schmidt, H., Are standards and regulations of organic farming moving away from small farmers knowledge? 5, Copyright (2005), with permission from The Haworth Press, Inc.

ecological responsibility in the entire production, processing, and distribution chain [10*].

A range of structural features and tactical management approaches are combined within whole farm systems. The design of a diverse crop rotation is the key to crop nutrition, weed, pest, and disease control. In Europe, yields of arable crops reach 60–80% of those in comparable conventional systems. In developing countries, organic farming practices increase crop yields with minimum external inputs. Lower variable costs and premium prices mean that organic-farming systems are profitable. Also, energy consumption and pollution are generally reduced in organic-farming systems [11**].

## 1.1.2 Development of organic farming

In the middle of the nineteenth century, agricultural practices started to change and after the setting of J. von Liebig's minimum theory, the first larger inputs of mineral substances (less soluble) in cultivated soil began. In the beginning of twentieth century, after establishing Haaber-Bosch synthesis in 1913 and industrial production, the first use of nitrogen mineral fertilizer began. After the introduction of urea synthesis from ammoniac in 1921 and production of the first herbicide (a side product of the military industry), "chemization" was introduced to agriculture; it accelerated after 1950 with the addition of modern farming equipment, rationalization, farm size growth, and the discovery of hybrids. Despite the first steps of organic farming taking place simultatneously with the new use of chemical substances in cultivated soil during the 1920s (Steiner, 1924), organic farming began to spread in the 1950s. In Switzerland and Austria, Müller introduced so-called organic-biological farming. In the 1960s, a book entitled *Silent Spring* contained revelations of major environmental problems. After the "green revolution" in the 1970s, it was mostly idealists with small farms and little farming experience who were involved in organic farming [12***].

Organic farming is the outcome of theory and practice that began in the early twentieth century, involving a variety of alternative methods of agricultural production, mainly in northern Europe. There have been three important movements [3]:

- Biodynamic agriculture, which appeared in Germany under the inspiration of Rudolf Steiner

* Reprinted from Food Quality and Preference, 12 (3), Torjusen, H., Lieblein G., Wandel, M., and Francis, C.A., Food system orientation and quality perception among consumers and producers of organic food in Hedmark County, Norway, 207, Copyright 2005, with permission from Elsevier.

** Reprinted from Advances in Agronomy, 70, Stockdale, E.A., Lampkin, N.H., Hovi, M., Keatinge, R., Lennartsson, E.K.M., Macdonald, D.W., Padel, S., Tattersall, F.H., Wolfe, M.S., and Watson, C.A., Agronomic and environmental implications of organic farming systems, 261, Copyright, 2005, with permission from Elsevier.

*** Reprinted from *Ekološko kmetijstvo/Organic Farming* (Knjižnica za pospeševanje kmetijstva), Bavec, M. (Ed.), 448 pp., Copyright 2005, with permission from Kmečki glas.

- Organic farming, which originated in England on the basis of the theories developed by Albert Howard in his Agricultural Testament (1940)
- Biological agriculture, which was developed in Switzerland by Hans-Peter Rusch and Hans Müller

Despite some differences in emphasis, the common feature of all these movements, which are the source of some of the terms protected by European Community rules, is to stress the essential link between farming and nature, and to promote respect for natural equilibrium. They distance themselves from the interventionist approach to farming, which maximizes yields through the use of various kinds of synthetic products. Despite the vitality of these movements, organic farming remained undeveloped in Europe for many years.

Throughout the 1950s, the main aim of farming was to achieve a major improvement in productivity so as to satisfy immediate needs for food and raise the European Community's rate of self-sufficiency. In those circumstances, organic farming was unlikely to be viewed very favorably.

By the end of the 1960s, however, and especially in the 1970s, organic farming came to the forefront in response to the emerging awareness of environmental conservation issues. New associations grew up, involving producers, consumers, and others interested in ecology and a lifestyle more in tune with nature. These organizations drew up their own specifications, with rules governing production methods. It was in the 1980s, however, that organic farming really took off, when the new production method continued to develop, along with consumer interest in its products, not only in most European countries, but also in the U.S., Canada, Australia, and Japan. There was a major increase in the number of producers, and new initiatives got under way for processing and marketing organic products.

This situation conducive to the development of organic farming was largely due to consumers' strong concern to be supplied with wholesome, environmentally friendly products. At the same time, the public authorities were gradually recognizing organic farming, including it among their research topics and adopting specific legislation (e.g., in Austria, France, and Denmark). Some member states also granted national or regional subsidies to organic farmers.

However, despite all these efforts, organic farming was still hampered by a lack of focus: consumers were not always sure about what was really covered by organic farming, and the restrictions it implied. The reasons for the confusion lay, among other things, in the existence of a number of different "schools" or "philosophies," the lack of harmonized terminology, the nonstandard presentation of products, and the tendency to blur the distinctions between concepts such as organic, natural, wholesome, and so on. The situation was not helped by cases of fraudulent labeling of organic methods. In the late 1980s, the development of official recognition and regulation of organic farming had started [3].

Adopting formal rules was the best way to give organic farming credibility in the quality products niche market. The European Community adopted a legal framework (Regulation [EEC] No. 2092/91) in the early 1990s. The movement toward official recognition of organic farming later spread to several other countries, and was followed by international initiatives. Because of its specific features, Regulation (EEC) No. 2078/922 provided further opportunities for the eligibility of financial support for organic farming.

- In November 1998, IFOAM adopted basic standards for organic farming and processing. The Federation, which was set up in 1972, brings together organizations from all over the world which are involved in organic production, the certification of products, research, education, and the promotion of organic farming. Although its specifications and standards are not binding, they do constitute very valuable guidelines, summarizing state-of-the-art methods of production, and processing of organic products.
- In June 1999, the Codex Alimentarius Commission adopted Guidelines for the Production, Processing, Labeling, and Marketing of Organically Produced Foods. These guidelines set out the principles of organic production from the farming stage through the preparation, storage, transport, labeling, and marketing of crop products. They are intended to enable member countries to draw up their own rules, on the basis of the principles, while taking account of specific national features.
- In 1999, the Food and Agriculture Organization (FAO) also embarked on an organic-farming work program, mainly concerned with promoting organic farming in developing countries.

Today IFOAM is a worldwide umbrella organization of the organic movement, uniting 771 member organizations in 108 countries. IFOAM's mission is to lead, unite, and assist the organic movement to its full potential. IFOAM's goal is the worldwide adoption of ecologically, socially, and economically sound systems that are based on the Principles of Organic Agriculture [13].

Within Europe, the development of organic agriculture took 30 years to occupy 1% of agricultural land and food markets. The recent food-safety crisis, however, resulted in an unforeseen growth in interest, wherein governments such as the United Kingdom are now targeting 30% of all agricultural land to become organic within the next 10 years!

In Argentina, the spectacular growth of organic land, from less than 500,000 hectares in 1999 to 3 million hectares in 2000, occurred mainly on grassland in response to organic meat demand (although this reflects reclassification and extension of certification rather than switches between farming systems). World meat-demand is increasing and is expected to continue to do so. If safety concerns continue to multiply (e.g., BSE, salmonella, dioxin), many countries, such as Japan (the largest importer of meat), are expected to seek more organic meat.

Future evidence of the health and environmental safety (or lack of it) of most transgenic foods will determine options for biotechnology and organic agriculture. For example, oil-crops production (especially soybean and rapeseed) is subject to major changes as oil crops are the focus of biotechnology development. At present, major organic soybean producers, such as Argentina, Brazil, and Paraguay, are being heavily confronted with GM soy and Bt corn that have become common in these countries. Consolidated knowledge on food safety of GM crops might either increase the potential of net exporting countries through improved production of GM foods or create new markets and exporters of organic commodities.

The current deregulation of agricultural policies leaves decisions on the most suitable type of agriculture and food production to market forces. Society's growing concern with the safety of food produced by conventional systems, as well as growing interest of multinational food enterprises and retailers in organic products, suggests that the growth of certified organic agriculture is likely to develop according to market rules.

However, moves toward the decoupling of agricultural support from production and the increasing emphasis on support to agriculture's role in providing public goods will also provide an impetus toward the adoption of environmentally friendly farming systems, including organic agriculture. Agro-environmental policies and global environmental conventions (especially those promoting carbon sequestration and biodiversity) are likely to trigger an increase of demand and supply for organic-agriculture products [5].

## 1.1.3   Standards

Organic farming systems are diverse and occur throughout the world. They are linked by common objectives of economic, environmental, and social sustainability. In many countries, organic farming now has a clear legislative basis and certification schemes for production and processing [11].

Governments have taken a major role in defining organic farming by creating legal standards. Many countries all over the world have established a certification and accreditation system in order to protect the justified expectations of consumers with regard to processing and controlling the product quality of organic goods and to protecting producers from fraudulent trade practices. As they are relevant to international trade, these standards do not only influence the organic farming movement on the national level but also have a converse impact across national borders. Organic farming was established in a "bottom-up" process as farmers aimed to design sustainable ways of using natural resources. Farmers' traditional knowledge and awareness of ecological, as well as social, affairs were the main base for the development of organic farming. Since public interest in organic farming has grown rapidly, the ownership of the process of defining organic farming is no longer in the hands of farmers and the original principles and aims of the movement seem to be threatened by a bureaucratic view of "recipe"-organic farming. However, unsolved problems also exist between the necessities of global harmonization and the local adaptability of the standards placed on organic farming [8].

The relationship between greater regulation and the diverse motivations and practices of organic farmers poses a dilemma for the organic movement. If consumers are to be certain that the products they buy are genuinely organic, and able to find out exactly what this means, then unless they know the producer directly, greater standardization seems inevitable. Therefore, there appear to be contradictory pressures on the organic movement, from outside and within. Some producers wish to retain small local organic markets where "food miles" are minimized, while others look to greater national and international coordination of standards and production methods [6].

As organic standards become established in an increasing number of countries, and as these standards become more coordinated and integrated, the degree to which the organic producer and organic consumer may be geographically separated grows. Furthermore, the trade in organic farm inputs may also grow, with organic producers having the option of buying mulch or organic fertilizers from distant sources. There may be doubts regarding the sustainability of the systems that have generated these purchased inputs. In addition, organic producers may be skeptical of such developments because they farm in this way to escape many aspects of the global trade in foodstuffs, and aim to supply local markets because of concerns about the energy efficiency of such a trade in organic products [6].

## 1.1.4 Some data about organic agriculture

Environmental awareness and food surplus in the EU are direct reasons for the significant increase of organic farming noted in the last two decades of the twentieth century; on the one hand, numerous environmental movements appeared in Europe (among consumers and producers), and on the other hand, farmers and governments were forced to deal with the food surplus and negative effects intensive farming left on the environment. Introduction of direct payments for environmentally friendly ways of farming increased interests for this type of production [14].

There are considerable differences among European countries, and growth in recent years depends on earlier years' conditions. Organic-farm growth stopped in Austria altogether; the share of organic farms stopped at 10% of all farms. Quick growth of high numbers of farms and cultivated surfaces included in organic farming ceased in Switzerland also, while other countries have only supported organic farming strongly in the last few years and have experienced real expansion—Italy, France, and Great Britain. Some countries are still in the initial stages—Greece and Portugal—while elsewhere governments decided that organic farming would be the mainstream agricultural policy, which resulted in fast growth in farms and processing plants (Denmark, Sweden) [12].

Organically cultivated areas in the EU are still expanding. In 1999, there were 2.9 million hectares of organically cultivated land, and at the end of 2002, there were 4.8 million hectares—a growth index of 161.

Organic farming is practiced in more than 100 countries, and more than 26 million hectares are managed organically worldwide. Of this total, 43.3% is in Australia, followed by Europe (23.8%) and Latin America (23.5%). The market for organic products is growing, not only in Europe and North America, but in many other regions, as well. In terms of the total amount of organic land, Australia leads the pack with 11.3 million hectares, followed by Argentina (2.8 million ha) and Italy (over 1 million ha). As most of the organic land area in Australia and Argentina is extensive grazing land, the global area dedicated to arable land is probably less than half. The proportion of organically, compared to conventionally, managed land, however, is the highest in Europe, and Latin America has the greatest total number of organic farms. The continued increase in organic land area is not just due to the ever greater interest in organic farming, but is also a result of improved access to information and data collection each time the study is updated. The area of certified "wild harvested plants," adds at least another 14.5 million hectares, according to various certification bodies [15]. In 2003, the market value of organic products worldwide reached 25 billion USD—the largest share of organic products being marketed in Europe and North America [15*].

Since the beginning of the 1990s, organic farming has rapidly developed in almost all European countries. Growth has, however, slowed down recently. In Europe almost 6.3 million hectares were managed organically by almost 170,000 farms. In the EU (including the new member states) almost 5.7 million hectares are under organic management, and there are more than 143,000 organic farms (as of December 31, 2003). This constitutes 3.4 percent of the agricultural area and 2% of the farms in the EU. A major development in the EU in 2004 was the launch of the European Action Plan for Organic Food and Farming. There are also substantial differences between individual countries regarding the importance of organic farming. More than 12% of agricultural land is organic in Austria, and 10% in Switzerland. Some countries have yet to reach 1%. The country with the highest number of farms and the greatest number of hectares is still Italy. Almost one fifth of the EU's organic land, and more than a quarter of its organic farms, are located there [15].

Organic farming is increasing in all new member states of the EU, which are mostly concerned with the exportation of organic foods to other countries within the EU (with exception of Slovenia, where almost all of the organic products generated are intended for domestic use, and the Czech republic). There are many initiatives in the EU to develop domestic markets for organic

---

* Reprinted with permission from IFOAM, Charles-de-gaulle-Str. 5, 53113 Bonn, Germany.

products in new member states to diminish the possibility of flooding the EU market with lower-priced organic products [14].

In North America, almost 1.5 million hectares are managed organically, representing approximately 0.3% of the total agricultural area. Currently, the total number of organic farms is more than 15,000. With the U.S. national standards [16] in place, the organic sector has been able to provide a guarantee to consumers that organic products using the new labeling process mean that specific practices were followed. The U.S. market has seen an increasing number of organic products being introduced, the number of certification agencies accredited by the United States Department of Agriculture (USDA) has grown, and talks are progressing to expedite the international trade of organic products. Since 1999, the Canadian industry has had a voluntary Canada Organic Standard that is not overseen by regulations. The organic industry continues to devote its energies toward implementation of mandatory national organic regulations to help expedite trade relations with such major trading partners as the U.S., the EU, and Japan. The North American market for organic products is reporting the highest growth worldwide. Organic food and drink sales were estimated to have expanded by 20% in 2003 to reach 10.8 billion USD. Consumer demand for organic products remains buoyant and the region is expected to account for the majority of global revenues in the foreseeable future [15].

Organic farming has been one of the fastest growing segments of U.S. agriculture for nearly a decade. Certified organic cropland for corn, soybeans, and other major crops more than doubled between 1992 and 1997, and doubled again between 1997 and 2001. Two organic livestock sectors—poultry and dairy—grew even faster. Farmers in 48 states dedicated 2.3 million acres of cropland and pastures to organic production systems in 2001. Over 1.3 million acres were used for growing crops. California, North Dakota, Minnesota, Wisconsin, Iowa, Montana, and Colorado had the most organic cropland. Colorado, Texas, and Montana had the largest amount of organic pastures and rangeland. Overall, certified organic cropland and pastures accounted for about 0.3% of total U.S. farmland in 2001. Only a small percentage of the top U.S. field crops—corn (0.1%), soybeans (0.2%), and wheat (0.3%)—were grown under certified organic farming systems. On the other hand, organic apples (3% of U.S. apple acreage), organic lettuce (5%), and other fruit and vegetable crops were more commonly organically grown in 2001. Markets for organic vegetables, fruits, and herbs have been developing for decades in the U.S. and fresh produce is still the top-selling organic category in retail sales [16]. The 2002 Census of Agriculture measured 11,998 certified-organic farms, representing 0.6% of all U.S. farming operations [2].

Organic farming is spreading rapidly in Latin America, Asia, Africa, and other less-developed regions, as well, but in most countries the products certified by European organizations for control are intended for export into developed countries and are not consumed by their own population. The development of organic farming is unique in Cuba, where trade embargos and international isolation triggered the management of biotic pests and diseases, improvement of organic-substance composting technology, and use of animal manure [14].

## 1.2   Organic food

The last few years have seen significantly increased interest in organic food—food grown using those husbandry principles and techniques that predated the introduction of modern agrochemicals and intensive farming methods. These principles are now applied with the benefit of modern scientific understanding and technologies to give a more sustainable system of food production. However, organic food production in the developed world is still dependent on fossil fuels for production, transport, and processing. Organic food is a small but growing sector of the food industry with an identity defined and protected by law. Its existence provides an element of consumer choice. Organic food production requires the same involvement of professional food scientists and technologists as the rest of the food industry and is subject to the same manufacturing practice and food safety regulations; however, it faces additional legal requirements regarding cultivation, composition, and labeling. Organic food is likely to contain lower residues of agricultural chemicals than its non-organic counterpart.

The use of animal waste as fertilizer, whether in producing organic or nonorganic food, needs to be properly managed. Even so, it may pose a risk of contamination with pathogenic microorganisms and consequent food poisoning from foods consumed without adequate, or any, cooking. In particular, fruit and salad vegetables meant for consumption without cooking, whether organic or nonorganic, should be thoroughly washed with potable water before consumption, and the public should be advised to do so by display notices and packaging.

Multivariate analyses have shown that traditional food quality aspects linked to food attributes that are directly observable through the eye or other senses like visible quality, freshness and taste, called "observation traits," are important to all consumers. In addition, those who purchase organic foods are more concerned with ethical, environmental, and health issues, called "reflection traits" linked to food attributes that are not directly observable like nutrient content, no use of GMOs, animal welfare. Three consumer orientations in the food market have been identified. Consumers with a "practical" orientation to the food market (looking for low prices, convenience through easy parking, broad selection of food) are less likely to buy organic food, while those with a "local" orientation (looking for locally produced food, support to local production) are more likely to buy it; "social" considerations (consumers are experiencing the shopping as pleasant, like receiving information through personal contact) are equally important to all. The results indicate that many interests of organic farmers coincide with concerns among those who buy organic food, and that these interests and concerns are more complex than the formal rules for organic agriculture. This may provide a basis for identifying common goals and improving communication and cooperation between consumers and producers of the organic food system [10*].

## 1.2.1 Organic food and legislation

Organic food rules are defined in Codex Alimentarius and IFOAM, and in the laws of other countries. Around the world, the same standards are applied to certified organic foods wherever organic farming legislation or standards based on IFOAM or Codex Alimentarious are used.

According to Regulation (EEC) No. 2092/91, organic food products contain organically produced ingredients and have certificates of organic production or processing. Organic food products may contain only 5% of conventionally produced agricultural ingredients listed in Annex VI part C. The use of additives and additional substances is strictly limited and must be avoided unless absolutely necessary; only the substances listed in Annex VI are allowed [14].

There are strict rules for the labeling and advertising of organic products in many countries. Regulation (EEC) No. 2092/91 states the following [3]:

- Labeling and advertising may refer to organic production methods only where they make it clear that the information relates to a method of agricultural production. The product concerned must comply with the provisions of Regulation (EEC) No. 2092/91. Moreover, the operator must be subject to the inspection measures laid down in the Regulation, and the name and/or code number of the inspection authority must be indicated. The rules on indications referring to organic production methods stipulate the minimum percentage of agricultural ingredients that must be of organic origin. Labeling and advertising of a food product may bear indications referring to organic production methods in the sales description only where at least 95% of the ingredients of agricultural origin are organic. Food products may thus contain up to 5% of ingredients produced by conventional methods as long as those ingredients are not available (e.g., exotic fruit) or are in very short supply on the Community organic market. Part C of Annex VI to Regulation (EEC) No. 2092/91 lists the ingredients concerned.
- Products with an organic content of 70% to 95% may bear indications referring to organic production methods only in the list of ingredients, but not in the sales description. Indications referring to organic production methods in the list of ingredients may not be more prominent than other indications in the list of ingredients. The percentage of ingredients of organic origin must be specified.

* Reprinted from *Food Quality and Preference*, 12 (3), Torjusen, H., Lieblein G., Wandel, M., and Francis, C.A., Food system orientation and quality perception among consumers and producers of organic food in Hedmark County, Norway, 207, Copyright 2005, with permission from Elsevier.

- Where the ingredients of organic origin represent less than 70% of the content of a product, the labeling and advertising may not bear any reference to organic production methods.
- Indications referring to conversion to organic production methods may be used; crop products that comply with the provisions of Regulation (EEC) No. 2092/91, and for which the operator is subject to inspection measures, may bear the words "product under conversion to organic farming," on condition that a conversion period of at least 12 months has been complied with before the harvest. The indications should not be such as to mislead the consumer. This faculty of referring to the conversion period is intended to help producers changing over to organic production in a period when the investment cost is usually heavy, by enabling them to enhance the product image after the first year.

Similar rules apply to the National Organic Program in the U.S. and many other non-European countries.

Food products can be marked as "organic" and carry a relevant label if a certificate was issued for them by an authorized control organization [14]. A certificate is issued under the following conditions:

- There were no GMOs used during processing.
- The food product or its ingredients were not submitted to ionic radiation.
- The food product contains at least 95% of organically produced agricultural ingredients and other ingredients of agricultural or nonagricultural origin listed in Annex VI Regulation (EEC) No. 2092/91.
- The food product contains not more than 5% of ingredients listed in Annex VI Regulation (EEC) No. 2092/91.
- Processing was strictly controlled.
- All accompanying documents are present in accordance with Regulation (EEC) No. 2092/91.

Lists of ingredients on organic food packaging must contain the source of ingredients from points 3 and 4 from the above list, in percentages. When selling to the consumer, the products may be marked as "organic," but the certificate has to be available for inspection. Food products containing less than 70% of organic ingredients can have organic ingredients listed on the packaging.

The list of ingredients should clearly state organic ingredients and the share or percentage with the text "food product with organic ingredients" presented in the same color, size, and font as all other ingredients. Food product in bulk must be accompanied by a certificate. On the basis of such certificate control, organizations can allow the use of the "organic" label for sale in retail.

Food product should not be marked as organic if:

- It does not comply with all provisions of Regulation (EEC) No. 2092/91 in all points.
- Its specification contains prohibited additives.
- Surfaces where free-growing plants are picked are not controlled and certified.
- It is processed in a conversion period and can be marked as "organic" only with the clear and visible statement "under conversion to organic farming," on the condition that a conversion period of at least 12 months has been complied with before the harvest, with production according to Regulation (EEC) No. 2092/91, and it contains only one ingredient of agricultural origin [14].

## 1.2.2  Quality of organic food

Consumers possess very high expectations for the quality of organic food. Usually, they expect natural and healthy food produced in the region, free of harmful compounds like pesticides, GMOs, or antibiotics.

Various countries show widely different attitudes toward organic production and different motivations for buying food products. In European countries, consumers reportedly perceive organic food products as healthier and more closely related to animal welfare and environment-friendly production. Many countries show an unbalanced development of supply and demand, and there is a very low market transparency connected to organic food production. The organic market is determined in part by development in the general food market [17]. Liberalization of the world market, EU enlargement, and GMOs are all factors that may boost the demand for organic product. A survey in Italy revealed that some consumers buy organic food based on expectations of better taste or higher quality, or because it is "trendy" and involves special products and niche production of "cleaner" (not preprocessed) food [18]. A German study showed that consumers search for product variety. This varies by household income; a single-male household consumes a significantly smaller number of different food products than does a household with children of the ages 7 to 17 years [19].

Dialogue and interaction between consumers, stakeholder organizations, companies, and the farming sector is essential for a mutual understanding and development of the organic food sector, and should be emphasized and catalyzed [20].

Studies comparing organic and conventional food have been conducted in recent decades. Different aspects have been studied: rests of pesticides, heavy metal content, nitrate content, content of vitamins and minerals, secondary metabolites, quality classes, storing, sensory properties, and so forth. Some researchers have done comprehensive literature studies, as well: Woese et al. [21] and Worthington et al. [22] for the period from 1926 to 1993; and Alföldi et al. [23] for the period between 1993 and 1998. Their studies reveal that, in most cases, there are no notable differences between organic and conventional products. In some cases, the organic products were shown to be better than the conventional ones, with regard to a higher content of vitamins and minerals and dry matter and a lower content of nitrates, rests of pesticides, and heavy metals. Only in a few cases were the conventional products better, e.g., with regard to quality classes or storage. These results show that organic food produced with low inputs has the same or higher quality as conventionally produced food with high inputs of fertilizers and pesticides.

Besides the usual laboratory methods, alternative methods such as bio-photon measurements, P-value, and feeding experiments have been developed and applied to check the quality of organic food compared to conventional products. Although these methods are still in development, the results are promising.

A comprehensive literature study on the potential benefits of organically produced food, based on more than 170 sources, confirmed that organic food is better than its conventional counterparts [24]. Study results confirmed that organic fruits and vegetables contain more vitamins (higher C-vitamin content was found in apples, cabbages, and tomatoes), more minerals, more secondary metabolites, higher dry matter content, fewer nitrates, fewer heavy metals, lower amounts of pesticide remains, and no ionic radiation. Consumers also found organic fruits and vegetables better tasting.

Organic cereals contain more essential amino acids, fewer mycotoxins, and considerably fewer pesticide and heavy metal remains than conventionally produced cereals; test animals also preferred organic cereals to conventional ones [25, 26]. Organic food products of animal origin have better fatty acid rations, and organic eggs contain more carotenes, fewer pesticide remains and antibiotics, and fewer aflatoxins. In addition to standard analytical procedures, new methods for determining food product quality are being developed [27–29], and their results are also in favor of organic food products.

There are many aspects of quality in organic food processing—one is the minimized use of additives. In organic farming, approximately 30 additives are allowed, compared to over 300 in conventional food processing (Table 1.1).

Organic food production is widely recognized as more friendly to the environment, better for animal welfare, and more controlled. The positive influences of organic food should be recommended for everyone, but especially for babies, pregnant and breast-feeding women, the elderly, the chronically ill, and vegetarians. Vegetarians may consume more vegetables than the average person, and those vegetables can contain too-high levels of carcinogenic substances when produced conventionally [32]. Pesticide contamination in conventional agriculture is over 60 times greater than in organic food. Organic products containing traces of pesticides contamination are found in only a minimal number of cases, which stem from the use of pesticides on neighboring conventional fields.

*Table 1.1* Number of Allowed Additives in Conventional and Organic Food [30*, 31]

| Additive | Conventional Food | Organic Food |
|---|---|---|
| Antioxidants | 55 | 11 |
| Colors | 48 | 1 |
| Gelling, thickening, and stabilizing agents | 74 | 12 |
| Flavoring agents | 19 | Not allowed |
| Preservatives | 50 | 3 |
| Acids | 28 | 6 |
| Sweeteners | 11 | Not allowed |
| Anticoagulants | 10 | Not allowed |

* Reprinted from *Ernte für das Leben*, Loczisky, T., Vergleich von Lebensmitteln aus konventionellem und biologischem Landbau, 22, Published in 1998, Copyright (2006), with permission from Bio Austria.

Conversely, there are hardly any samples of fruits and vegetables from conventional production without pesticide contamination. In an alarming number of cases, products are even sold with pesticide contents exceeding the legal limits, according to the Chemical and Veterinary Inspectorates Baden-Württemberg [33].

Avoiding pesticides and additives in food is often the consumer's motivation for buying organic food, but taste is equally important. The introduction of alternative field crops in organic crop rotation provides the possibility to produce new and better-tasting organic foods with higher nutritional value.

# 1.3  Multifunctionality of organic agriculture

## 1.3.1  Sociological aspects

These days, agriculture has many other functions in addition to food production and economics, such as maintaining the cultural landscape. Among the different methods of agricultural production, organic farming integrates different approaches that are oriented toward the improvement of human health, the preservation of workplaces, the cultivation of small farms, and so forth. With its high-quality products, organic agriculture works to maintain inhabited rural areas, preserve biodiversity, soil fertility, and diminish the negative influences of agriculture upon the environment.

The reasons for converting from conventional or traditional farming practices to organic farming vary; they may be economical, environmental, philosophical, and so forth. Organic farming is a way of thinking and a way of living. It provides an economic alternative to farms that cannot compete with the decreasing prices of conventional products on the world market. Organic farming provides the possibility to produce healthy food for the market; it is a way of farming that is adapted to the needs of inhabitants without a surplus of food or pollution costs; and it results in premium-priced products. In many countries, additional financial support for organic farming also exists [12].

Over a million farmers are involved in organic agriculture worldwide, growing the full spectrum of agricultural products in organic quality. Each and every organic farmer is making a positive contribution to the improvement of the environment and the enhancement of community life. Organic farmers and growers understand social, economic, and environmental relationships and the historical development of agriculture. They know that farming is hard work, and that dumping toxic pesticides and fertilizers on fields and loading animals with antibiotics simply is not the answer for a sustainable and healthy future. Organic farmers use innovative and ingenious practices to enhance the traditional knowledge they have received from thousands of generations of farmers [13].

## 1.3.2   Biodiversity

Biodiversity performs a variety of ecological services beyond the production of food, including the recycling of nutrients, the regulation of microclimate and local hydrological processes, the suppression of undesirable organisms, and the detoxification of noxious chemicals. Because biodiversity-mediated renewal processes and ecological services are largely biological, it is argued that their persistence depends upon the maintenance of biological integrity and diversity in agroecosystems. When these natural services are lost due to biological simplification, the economic and environmental costs can be quite significant. Economically, in agriculture the burdens include the need to supply crops with costly external inputs, because agroecosystems deprived of basic regulating functional components lack the capacity to sponsor their own soil fertility and pest regulation. Often the costs involve a reduction in the quality of life due to decreased soil, water, and food quality when pesticide and/or nitrate contamination occurs [34].

The net result of biodiversity simplification for agricultural purposes is an artificial ecosystem that requires constant human intervention, whereas in natural ecosystems the internal regulation of function is a product of plant biodiversity through flows of energy and nutrients, and this form of control is progressively lost under agricultural intensification [35*].

The intensification and expansion of modern agriculture is among the greatest current threats to worldwide biodiversity. Over the last quarter of the twentieth century, dramatic declines in both range and abundance of many species associated with farmland have been reported in Europe, leading to growing concern over the sustainability of current intensive farming practices. Purportedly "sustainable" farming systems, such as organic farming, are now seen by many as a potential solution to this continued loss of

* Biodiversity and Ecosystem Function, 1993, 57, Biodiversity and ecosystem function in agro-ecosystems, Schulze, E., and Mooney, H.A., Eds., Swift, M.J. and Anderson, J.M., Copyright 2005, with kind permission of Springer Science and Business Media.

biodiversity and receive substantial support in the form of subsidy payments through EU and national government legislation [36].

Among transitional countries in Europe after 1990, the main agricultural production has become oriented to the market and is based on modern Western European technology. This production is associated with increasing pollution and has become a serious threat to biodiversity. Many wild plants are endangered due to genetic erosion within species. The traditional crops and varieties are being replaced by imported materials, and the use of chemicals has been increasing. Many of the traditional varieties have been neglected or lost. The existing germplasm collections are incomplete and frequently do not include traditional varieties. Legislation is not efficient enough to protect the environment and biotic diversity. School, media, religious, sport, tourist, and other organizations need to be more active in promoting respect for biodiversity [37].

Organic farming systems have larger and more diverse flora, insect, and bird populations, caused by changes in both habitat structure and field management [11].

In most of the studies presented in the Soil Association Biodiversity report, important differences were found between the biodiversity on organic and conventional farms, with substantially greater levels of both abundance and diversity of species generally found on the organic farms:

- Plants: 5 times as many wild plants in arable fields, 57% more species, and several rare and declining wild arable species found only on the organic farms, including some Biodiversity Action Plan species.
- Birds: 25% more birds at the field edge, 44% more in-field in autumn/ winter; 2.2 times as many breeding skylarks and higher skylark breeding rates.
- Invertebrates: 1.6 times as many of the arthropods that comprise bird food; three times as many non-pest butterflies in the crop areas; one to five times as many spider numbers and one to two times as many spider species.
- Crop pests: significant decrease in aphid numbers; no change in numbers of pest butterflies.
- Distribution of the biodiversity benefits: though the field boundaries had the highest levels of wildlife, the highest increases were found in the cropped areas of the fields.
- Quality of the habitats: both the field boundary and crop habitats were more favorable on the organic farms.
- Field boundaries: had more trees, larger hedges and no spray drift; crops were sparser, with no herbicides, allowing more weeds; more grassland and a greater variety of crop types.
- Organic farming: identified as having many beneficial practices, reversing the trends in conventional farming that have caused the decline in biodiversity; crop rotations with grass leys, mixed spring

and autumn sowing, more permanent pasture, no use of herbicides or synthetic pesticides and use of green manuring [38].

Hole et al.* [36] reviewed 76 studies on different aspects of biodiversity on organic and conventional farming systems published over the last 20 years, which demonstrated that species abundance or richness, across a wide range of taxa, tend to be higher on organic farms than on locally representative conventional farms. Of particular importance from a conservation perspective is that many of these differences apply to species known to have experienced declines in range or abundance as a consequence of past agricultural intensification, a significant number of which are now the subject of direct conservation legislation (e.g., skylark, lapwing, greater and lesser horseshoe bat, corn buttercup [*Ranunculus arvensis*], and red hemp-nettle are all U.K. government Biodiversity Action Plan species). These biodiversity benefits are likely to derive from the specific management practices employed within organic systems that are either absent or only rarely utilized in the majority of conventional systems.

Predatory carabides, staphylinids, and spiders were monitored concerning the development of arable farming systems in the Netherlands. During a six-year period, epigeic predators in the conventional, integrated, and organic systems were sampled using pitfall traps. The effects of crop type and farming system on species diversity, abundance, and guild structure were analyzed using trap data from wheat, pea, sugar beet, potato, onion, and carrot fields. Predator abundance and species composition were found to be clearly affected by the farming system [39].

## 1.3.3 Environmental impacts

A key policy objective of common agriculture policy (CAP) is the sustainability of agriculture and environment, where sustainable development must encompass food production alongside conservation of finite resources and protection of the natural environment so that the needs of people living today can be met without compromising the ability of future generations to meet their own needs. This objective requires farmers to consider the effect of their activities on the future of agriculture and how today's systems shape the environment. As a consequence, farmers, consumers, and policy makers have shown a renewed interest in organic farming. Farmers in Europe are now expected to respect basic environmental standards without any financial compensation, and the "polluter pays" principle is being applied. Which means whoever causes pollution is responsible for the cost of repairing any damage. However, the agro-environment measures under the rural development programs offer farmers payments for environmental commit-

---

* Reprinted from *Biological Conservation*, 122 (1), Hole, D.G., Perkins, A.J., Wilson, J.D., Alexander I.H., Grice, F., and Evans, A.D., Does organic farming benefit biodiversity? 113, Copyright 2005, with permission from Elsevier.

ments that go beyond good farming practice. Organic farmers are entitled to claim agro-environmental premiums, since it is recognized that this particular farming system benefits the environment. Organic farming can also be promoted through investment aid in the areas of primary production, processing, and marketing. With all of these provisions in place, the framework of rural development policy is likely to contribute considerably to the expansion of organic farming. To understand the role and operation of organic farming within EU farm policy, it needs to be seen within a range of different contexts, including consumers' concerns, quality assurance and regulation, the extent of organic farming in the EU today, and the role of integrated farming, organic farming, and rural development [40].

Organic farming systems have larger and more diverse flora, insect, and bird populations, caused by changes in both habitat structure and field management [11]. Organic farming systems perform better than integrated and conventional systems and CFSs with respect to nitrogen loss, pesticide risk, herbaceous plant biodiversity, and most of the other environmental indicators [41*]. If organic farming is to be truly sustainable, it must maintain levels of soil fertility sufficient for economic crop production in the long-term, while also protecting the environment [42**].

Organic farming enhances soil structures, conserves water, and ensures the conservation and sustainable use of biodiversity. Organic agriculture dramatically reduces external inputs by refraining from the use of synthetic fertilizers and pesticides, genetically modified organisms, and pharmaceuticals. Pests and diseases are controlled with naturally occurring means and substances according to traditional and modern scientific knowledge, increasing both agricultural yields and resistance to diseases. Organic agriculture adheres to globally accepted principles, which are implemented within local socioeconomic, climatic, and cultural settings. Agricultural contaminants such as inorganic fertilizers, herbicides, and insecticides from conventional agriculture are a major concern all over the world. Eutrophication—the suffocation of aquatic plants and animals due to the rapid growth of algae, referred to as "algae blooms"—is killing lakes, rivers, and other bodies of water. The damage from persistent herbicides and insecticides can extend beyond target weeds and insects when introduced into aquatic environments. These chemicals have accumulated up the food chain, whereby top predators often consume toxic dosages. Organic agriculture restores the environmental balance and has none of these or other such deleterious effects on the environment [13].

* Reprinted from *Agric., Ecosyst. Eviron.*, 95 (1), Pacini, C., Wossink, A., Giesen, G., Vazzana, C. and Huirne, R., Evaluation of sustainability of organic, integrated and conventional farming systems: a farm and field-scale analysis, 273, Copyright 2005, with permission from Elsevier.
** Reprinted from *Agric., Ecosyst. Environ.*, 105 (1-2), Gosling, P. and Shepherd, M., Long-term changes in soil fertility in organic arable farming systems in England, with particular reference to phosphorus and potassium, 425, Copyright 2005, with permission from Elsevier.

# 1.4  Organic crop production

The keystone of organic farming and growing is creating nutrient-rich, healthy soil that contains abundant organic matter, which helps to prevent erosion, retain water, and control the release of nutrients to plants. Organic farmers use a variety of methods to maintain and build soil fertility. By utilizing compost and recycling plant and animal waste materials, organic farmers create high-quality organic matter that is rich in beneficial microorganisms. Organic farmers also use cover crops to rejuvenate the soil and prevent erosion. Complex crop rotation—the planting of a diverse variety of crops on a field over time—is another integral method that organic farmers use to create a sustainable and efficient farming system [13].

## 1.4.1  Rules in organic crop production

Basic rules of organic crop production are written in standards and legislations. As this book is written for European and North American conditions, their legislations for organic crop production are presented as an example of the basic production principles.

The National Organic Program (NOP) in the U.S. provides the following rules for plant production [43]:

- A producer of an organic crop must manage soil fertility, including tillage and cultivation practices, in a manner that maintains or improves the physical, chemical, and biological condition of the soil and minimizes soil erosion. The producer must manage crop nutrients and soil fertility through rotations, cover crops, and the application of plant and animal materials. The producer must manage plant and animal materials to maintain or improve soil organic matter content in a manner that does not contribute to contamination of crops, soil, or water by plant nutrients, pathogenic organisms, heavy metals, or residues of prohibited substances. Plant and animal materials include raw animal manure, composted plant and animal materials, and uncomposted plant materials. Raw animal manure must either be composted, applied to land used for a crop not intended for human consumption, or incorporated into the soil at least 90 days before harvesting an edible product that does not come into contact with the soil or soil particles and at least 120 days before harvesting an edible product that does come into contact with the soil or soil particles. Composted plant or animal materials must be produced through a process that establishes an initial carbon-to-nitrogen (C:N) ratio of between 25:1 and 40:1 and achieves a temperature between 55°C and 76°C. Composting operations that utilize an in-vessel or static aerated pile system must maintain a temperature within that range for a minimum of 3 days. Composting operations that utilize a windrow composting system must maintain a temperature within that range for a minimum of 15 days, during which time the materials must be turned five times.

- In addition to these practices and materials, a producer may apply a crop nutrient or soil amendment included on the National List of synthetic substances allowed in crop production. The producer may apply a mined substance of low solubility. A mined substance of high solubility may only be applied if the substance is used in compliance with the annotation on the National List of nonsynthetic materials prohibited in crop production. Ashes of untreated plant or animal materials which have not been combined with a prohibited substance and which are not included on the National List of nonsynthetic substances prohibited for use in organic crop production may be used to produce an organic crop. A plant or animal material that has been chemically altered by a manufacturing process may be used only if it is included on the National List of synthetic substances allowed for use in organic production. The producer may not use any fertilizer or composted plant and animal material that contain a synthetic substance not allowed for crop production on the National List or use sewage sludge. Burning crop residues as a means of disposal is prohibited, except that burning may be used to suppress the spread of disease or to stimulate seed germination.

- The producer must use organically grown seeds, annual seedlings, and planting stock. The producer may use untreated nonorganic seeds and planting stock when equivalent organic varieties are not commercially available, except that organic seed must be used for the production of edible sprouts. Seed and planting stock treated with substances that appear on the National List may be used when an organically produced or untreated variety is not commercially available. Nonorganically produced annual seedlings may be used when a temporary variance has been established due to damage caused by unavoidable business interruption, such as fire, flood, or frost. Planting stock used to produce a perennial crop may be sold as organically produced planting stock after it has been maintained under a system of organic management for at least 1 year. Seeds, annual seedlings, and planting stock treated with prohibited substances may be used to produce an organic crop when the application of the substance is a requirement of Federal or State phytosanitary regulations.

- The producer is required to implement a crop rotation, including but not limited to sod, cover crops, green manure crops, and catch crops. The crop rotation must maintain or improve soil organic matter content, provide for effective pest management in perennial crops, manage deficient or excess plant nutrients, and control erosion to the extent that these functions are applicable to the operation.

- The producer must use preventive practices to manage crop pests, weeds, and diseases, including but not limited to crop rotation, soil and crop nutrient management, sanitation measures, and cultural practices that enhance crop health. Such cultural practices include the selection of plant species and varieties with regard to suitability to site-specific conditions and resistance to prevalent pests, weeds, and diseases. Mechanical and biological methods that do not entail application of synthetic substances may be used as needed to control pest, weed, and disease problems that may occur. Pest control practices include augmentation or introduction of pest predators or parasites; development of habitat for natural enemies; and nonsynthetic controls such as lures, traps, and repellents. Weed management practices include mulching with fully biodegradable materials; mowing; livestock grazing; hand weeding and mechanical cultivation; flame, heat, or electrical techniques; and plastic or other synthetic mulches, provided that they are removed from the field at the end of the growing or harvest season. Disease problems may be controlled through management practices which suppress the spread of disease organisms and the application of nonsynthetic biological, botanical, or mineral inputs. When these practices are insufficient to prevent or control crop pests, weeds, and diseases, a biological or botanical substance or a synthetic substance that is allowed on the National List may be used provided that the conditions for using the substance are documented in the organic system plan. The producer must not use lumber treated with arsenate or other prohibited materials for new installations or replacement purposes that comes into contact with soil or livestock.

In the European Union, organic plant production is regulated in the Council Regulation (EEC) No. 2092/91 in part A of Annex I, as follows [3]:

- The fertility and the biological activity of the soil must be maintained or increased by the cultivation of legumes, green manure, or deep-rooting plants in a multi-annual rotation. By-products from organic livestock farming may also be used within the limits set out in part B of Annex I (170 kg of nitrogen per year and per hectare), as may organic material, composted or not, from holdings producing according to organic methods.
- Where this is insufficient to ensure adequate crop nutrition or soil conditioning, further means are required. However, no organic or mineral fertilizers may be used other than those listed in part A of Annex II to the Regulation, which mainly concerns natural, relatively insoluble minerals obtained by means other than chemical synthesis.
- Microorganism preparations (not genetically modified) may be used to improve the general state of the soil or the availability of nutrients in the soil or the crop, where such a need has been recognized by the Member State concerned.

- Crop protection against pests and disease, and weed control, must be achieved as far as possible without the use of plant health products. Crop protection is primarily achieved through the choice of naturally resistant species and varieties, an appropriate rotation program, mechanical cultivation procedures, flame weeding, and the protection of natural enemies of pests (e.g., care of hedges, nests, etc.).
- In cases of immediate threat to the crop, the plant health products listed in part B of Annex II to the Regulation may be used under certain conditions. The list comprises four categories of product: certain products of animal or plant origin, products based on microorganisms, certain substances that must be used in traps or distributors, and certain substances traditionally used in organic farming prior to the adoption of Regulation (EEC) No. 2092/91.
- The minimum conversion period for the change from conventional to organic farming is 2 years (before sowing) for annual crops, and 3 years (before the first harvest) for perennial crops other than grassland. In certain cases the conversion period may be extended or reduced, having regard to previous parcel use. The Member States establish the conditions for increasing or reducing the conversion period.

Different organic crop production practices in varying climatic conditions have been developed, based on organic agriculture legislation and IFOAM, Codex Alimentarius, and several organic farmers' association (Demeter, Bioland, Naturland, Biodar, BioErnte, etc.) standards. These practices vary from conventional, integrated, and conservation agriculture. Organic agriculture and crop production has become the topic of many studies in recent years; some of the findings are presented in the next chapter.

### 1.4.2   Organic crop management

Organic farming is founded on the idea that soils with sufficient organic matter content, good structure, and rich and variegated living organisms can provide a base for healthy crops. Soil cultivation is, therefore, a central topic of organic farming. Plant nutrient supply is based on the previous input of organic matter in the soil, which is decomposed and mineralized by soil organisms.

Nutrient management in organic systems is based on fertility-building leys that fix atmospheric nitrogen (N), combined with the recycling of nutrients via bulky organic materials, such as farmyard manure (FYM) and crop residues, with only limited inputs of permitted fertilizers [42].

Conventional agriculture in industrialized countries is based on specialized agroecosystems, where high yields are obtained from sequences of annual crops (cash crops) with large inputs of chemicals. A sustainable agriculture should rely on the major exploitation of native resources, e.g., sun energy, water from precipitations, atmospheric nitrogen, and soil organic matter. Crop rotation, based on the inclusion of polyannual legumes, is one of the most powerful management practices for pursuing such aims due to its implications for maintaining soil fertility, saving energy, and avoiding pollution [44].

### 1.4.2.1    Influence of precrop

Plant nutrient supply for organic cash crops stipulates that cropping during a 2-year conversion period must maintain or increase soil fertility and leguminous green manure crops, such as red clover (*Trifolium repens* L.). The goals of such a "fertility-building" conversion strategy include the provision of nutrients (especially nitrogen) to support subsequent cropping; to improve the physical, biological, and chemical attributes of soil that are associated with fertility; and to suppress weeds, pests, and diseases, in order to provide a "clean" seedbed. The loss of income associated with growing a green manure crop for 2 years, however, may be too great for many growers to accept. Growers could consider cash cropping during the conversion period to generate some income, as an alternative to this strategy, with crops grown in the second year of conversion possibly benefiting from "in-conversion" price premiums. Cash cropping during the conversion period has the potential to raise farm incomes compared with green manure crops, but may reduce the yield of the subsequent organic crops. Furthermore, the sustainability of the organic rotation may be affected by the choice of conversion cropping: strategies that do not comprise a fertility-building phase may lead to declining yields over time due to imbalances in nutrient budgets, deterioration of soil structure and other aspects of soil fertility, and a buildup of weeds, pests, and diseases [45].

Investigations into the residual effect on the succeeding crop (sunflower) of a long-term rotation, which included a 4-year lucerne ley, in comparison with a sequence of annual crops, was conducted in a Mediterranean environment (central Italy). Similar yields were obtained in contrasting conditions. Up to 3000 kg ha$^{-1}$ seed yield and 1400 kg ha$^{-1}$ oil yield, approximately, were recorded either with lucerne as the preceding crop, without any addition of mineral nitrogen (Nmin) and with mechanical weed control, or with safflower as the preceding crop, supplying sunflower with 130 kg ha$^{-1}$ of Nmin and chemical weed control [44].

One year's green manure crop cultivation could form the base for supplying nitrogen in an all-arable or "stockless" crop rotation in organic farming systems. Three 4-year crop rotations from two long-term field trials on stockless organic agriculture in the U.K. and Germany were analyzed using data collected over 5 and 7 years, respectively. All three rotations were based on one year's clover green manure with the aim to assess the N-supply by the green manure and its influence on the succeeding crops. The estimated N-input by the symbiotically fixed N from the clover ranged from 20 to 350 kg N ha$^{-1}$. The average for two of the three rotations demonstrated that N-input was greater than N-export. In the third case, N-supply was much lower than N-export, due to weather- and pest-related damage to the clover.

In the individual rotations, the influence of varying green manure N-accumulation on N-uptake of the cash crops was not significant. However, combining the rotations by using partial correlation analysis revealed significant correlations in some cases [46].

### 1.4.2.2 Intercropping

Crop rotations, sequences, and space in the form of cover crops, agroforestry, crop/livestock mixtures, and so forth can enhance diversity through intercropping. Correct biodiversification results in pest regulation through restoration of natural control of insect pests, diseases, and nematodes, and also produces optimal nutrient recycling and soil conservation by activating soil biota, all factors leading to sustainable yields, energy conservation, and less dependence on external inputs. Diversification can also take place outside the farm, e.g., in crop-field boundaries with windbreaks, shelterbelts, and living fences, which can improve habitat for wildlife and beneficial insects, provide sources of wood, organic matter, resources for pollinating bees, and, in addition, modify wind speed and the microclimate [34].

Recent studies of mixed intercropping of spring barley (*Hordeum vulgare* L.) with field pea (*Pisum sativum* L.), faba bean (*Vicia faba* var. *minor* L.), or narrow-leafed lupin (*Lupinus angustifolius* L.) compared with sole cropping in field experiments on two types of soils showed that grain legumes were dominant in intercrops on the sandy loam soil, except for lupin, whereas barley was dominant in intercrops on the sandy soil site. Combined intercrop grain yields were comparable to grain yields of the respective sole-cropped grain legume or sole-cropped, fertilized barley on each soil site. On the sandy loam soil, pea-barley and faba bean-barley intercrops increased the proportion of plant N derived from nitrogen fixation in grain legumes and increased the barley grain N concentration (from 1-7 to 2-2 mg g$^{-1}$) compared with sole cropping. Lupin-barley intercrops did not show intercropping advantages to the same degree as faba bean and pea, but lupin constituted a more stable yield proportion of the combined intercrop yield over locations [47].

Intercropped crops represent an important production system in organic farming, especially maize/climbing bean mixture, due to its high content of protein in bean seeds for human diet and silage production for ruminants [48].

Stand density and crop uniformity are very important for natural weed control. If the plant density is reduced, the dry matter production of the weeds will increase due to higher light intensity at ground level. The long-term experiment determined how long it took various crops to reach this density during their vegetation phases. If this period is known, the crop will only need to be protected with other weed control measures before and after this date, if at all. In creating crop rotations, weed problems could be reduced based on weed habits. Critical limit was found to be 50% light reduction, which caused a significant decline in weed infestation [49].

Described alternative crops are subject to attack by wide range of insect pests and diseases, but there are considerable variations in the relative importance of various pests and diseases in different growing conditions.

Technical, economical, and environmental factors are forcing organic farmers to adopt new sustainable methods, such as the use of microbial antagonists for the control of soilborne pathogens. Recent research has mainly focused on antagonistic fungi and bacteria, often not providing consistent or satisfying results. Different biocontrol strategies should be developed for different pathogens. The use of microorganisms can play an important role in a more complex vision of crop protection and alternative crops. The production system, formulation, and methods of application are crucial to maintaining and improving the efficacy of microbial antagonists [50].

## 1.5   Contribution of organic agriculture to the conservation of alternative crops and their utilization development

Intensive agricultural landscapes are planted mostly with a single or a few crop species and a handful of varieties or only hybrids (for example, about 95% of genetically modified maize in the Corn Belt is produced as a monoculture of maize), or cereal crop rotation, 2-year, or two-plant crop rotation. No more than 70 plant species are spread over approximately 1440 million hectares of cultivated land in the world today, and a number of cultivated species of major crops are decreasing continually in developed countries.

Biodiversity is not foreign to traditional farmers in the Third World. Traditional cropping systems contain numerous varieties of domesticated crop species, as well as their wild relatives. The species richness of all biotic components of traditional agroecosystems is comparable to that of many natural ecosystems. Organic farming based on traditional farming systems offers a means of promoting diversity of diet and income, minimization of risk, reduced insect and disease incidence, efficient use of labor, intensification of production with limited resources, maximization of returns, and stability of production under low levels of technology [52]. Many scientists have recognized that traditional farming systems can be models of efficiency, as they incorporate careful management of soil, water, nutrients, and biological resources [53]. Biodiversity performs key ecological services and can lead to agroecosystems capable of sponsoring their own soil fertility, crop protection, and productivity. Diversity can be enhanced over time through crop rotations and sequences and over space in the form of cover crops, intercropping, agroforestry, crop/livestock mixtures, and so forth. Correct biodiversification results in pest regulation through the restoration of natural control of insect pests, diseases, and nematodes and also produces optimal nutrient recycling and soil conservation by activating soil biota, all factors leading to sustainable yields, energy conservation, and lesser dependence on external inputs [34].

Due to the monotone technological development of growing only a few crops, numerous scientists have conducted research and development over recent decades in the areas of selection, production, and use of previously less-produced species of crops. These plants are usually called rare, alternative, neglected, disregarded, and new crops; new crops also include the category of genetically modified crops.

Studies of the less-produced crops were first supported in the 1980s when a surplus of main crops appeared on the market. New directions have been undertaken through the introduction of uncultivated fields (set aside), "agricultural" forests, the production of biomass for energy, the allotment of recreational functions to fields, the permanent establishment of "close-to-nature" areas, and the production of farm crops intended for industrial use. The most successful activities in Europe have been the production of economically popular crops and the marketing of products, semi-products, and final organic food products in different ways. The use of rare crops has resulted in product competitiveness, rich nutritional value, tradition, locality, special quality, and biodiversity in increased closeness to nature. The continued use of "new" plants depends most heavily on the adaptation capabilities of plants, market attraction for producers, and industry interest.

Some crops have been forgotten over time due to economical or religious influences. It will be a great waste if scientists forget all about rare crops and disregard new plants. The world has realized that poor crop biodiversity, monocultures, negative influences on farming, market surplus, extreme variations of prices, and the poor nutritional value of plants cause great damage to long-term closeness to nature and the survival of humankind.

Rich food sources of amino acids, such as amaranth, quinoa, and some legumes, were often being replaced by monotone foods. In some cases, it was difficult to change their content by breeding new like corn cultivars "Opaque 2." Due to the reduction of crop biodiversity and a mostly uniform use of natural resources, ecological oversensitivity to pests and diseases was increasing [54].

The inclusion of rare crops must become a rule in organic farming. Rare crops bring diversity into crop rotation and provide new possibilities for soil cultivation. Crop diversity can influence the occurrence of weeds, diseases, and soil erosion. The introduction of new plants makes local communities more independent and reduces transport expenses. Many plants could become useful in the latex industry, the rubber industry, starch production, vegetable oil production, and so forth. Some alternative fiber plants also appear very promising.

New plants play an important role in human and animal nutrition. Some have found a special place in medicine—for example, *Catharanthus (Vinca) roseus*, which contains anticarcinogenic substances such as vinblastin and vinkristin, or *Taxus breviflora*, which contains the active substance taxolreg (paclitaxel). Grain amaranths can prevent blood anemia and pumpkin oil can help with prostate problems. The effects of natural resources on the immune system in cases of HIV infections have also been studied [55].*

* Reprinted from *Diversifying U.S. crop production*, in Progress in new crops, Janick, J. et al., 1998, Copyright 2005 with permission from ASHS.

In this book, we deal with crops that could bring diversity into crop rotation with some entrepreneurship. Without new research or, traditional crops and related culinary arts will soon be forgotten. Some traditional crops are already disappearing from our fields. Special attention must be paid to neglected crops, such as flax and hemp, and also to the crops that could become "new" crops (amaranths) in moderate continental environments or the crops that do not grow well in continental environments (peanuts, chickpea) but can still be grown by enthusiasts.

Knowledge about these plants is always welcome, especially with relation to climatic changes. From history, we are familiar with the cereals that have disappeared from primary cereal centers and Africa due to changes in climate. Without the threat of direct greenhouse effect, new dry areas in Canada and the U.S. would appear in intensive crop production as a consequence of gathering underground water for irrigation purposes. The choice of potential rare crop species in organic agriculture depends on climatic circumstances, soil characteristics, production possibilities, tradition, needs of special food quality, and more. Potential alternative crops for organic crop production and better crop rotation are [56]:

- Buckwheat (*Fagopyrum esculentum* Moench) (syn.: *F. sagittatum* Gilib., *F. vulgare* T. Nees, Hill)
- Foxtail, foxtail millet, Italian millet, German millet, Hungarian millet, Siberian millet (*Setaria italica* [L.] P. Beauv. ssp. *maxima* Alef.), (syn.: *Panicum italicum* L.)
- Oil (seed) pumpkins, oil squash, oil marrow, oilbearing gourd (*Cucurbita pepo* L. convar. *citrullina* [L.] Greb. var. *styriaca* Greb.) (syn.: *C. pepo* L. convar. *giromontiina* [L.] Greb. var. *oleifera* Pietsch.)
- Chickpea, garbanzo, bengal gram, egyptian pea (*Cicer arietinum* L.), (syn.: *C. edessanum* Stapf., *C. grossum* Salisb., *C. sativum* Schkuhr)
- White mustard, yellow mustard (*Sinapis alba* L.) (syn.: *Brassica alba* [L.] Boiss., *Raphanus albus* Crantz)
- Industrial hemp (*Cannabis sativa* L.) (syn.: *C. macrosperma* Stokes, *C. lupulus* Scop.)
- Flax (cultivated) (*Linum usitatissimum* L.)
- Poppy, garden poppy (*Papaver somniferum* L.) (syn.: *P. officinale* Gmel., *P. album* Crantz, *P. hortense* Hussekot.)
- Proso millet, common millet, hog millet, broomcorn millet, Russian millet, brown corn, Indian millet (*Panicum miliaceum* L.) (syn.: *Milium panicum* Mill., *M. esculentum* Moench, itd.)
- Oil seed rape, rape, colza (*Brassica napus* L. emend. Metzg var. *napus*)
- Spelt, large spelt, dinkel, dinkel wheat, german wheat (*Triticum spelta* L.) (syn.: *T. zea* Host, *T. vulgare* Vill. var. *spelta* Alef., *T. arduini* Mazz.)
- Small spelt, einkorn (wheat), one-grained wheat (*Triticum monococcum* L.) (syn.: *Triticum vulgare bidens* Alef., *Nivieria monococcum* Ser.)

- Emmer (wheat), two-grained spelt (*Triticum dicoccon* Schrank) (syn.: *T. dicoccum* Schuebl. *T. turgidum* L. var. *dicoccon* (Schrank) Thell., *Triticum spelta*)
- Kamut (wheat) (*Triticum turgidum* ssp. *polonicum*)
- Camelina, conio, gold of pleasure, false flax (*Camelina sativa* [L.] Crantz) (syn.: *C. glabrata* [DC.] Fritsch, *C. sativum* ssp. *sativa* Thell., *Myagrum sativum* L., *Alyssum sativum* Scop.)
- Sweet potato, batate, ipomea batata (*Ipomoea batatas* [L.] Lam., Poir.) (syn.: *Convolvolus tuberosus* Vell., *Batatas edulis* Choisy)
- Sunflower (*Helianthus annuus* L.) (syn.: *H. indicus* L., *H. grandiflorus* Wender., *H. cultus* Vencl.)
- Grain amaranths: a) bush greens, red amaranth, huautli (*Amaranthus cruentus* L., syn.: *A. paniculatus* L.); b) prince's feather (*Amaranthus hypochondriacus* L.); c) love-liebleeding, Inca wheat, cat-tail, tumbleweed, Omca wheat, Quihuicha, tassel flower (*Amarathus caudatus* L.)
- Triticale (*Triticosecale* Wittm. & Camus) (syn.: *T. rimpaui* Wittm., *Triticale* Müntzing)
- Jerusalem arthicoke (*Helianthus tuberosus* L.) (syn.: *H. mollissimus* E. Wats)
- Groundnut, peanut (*Arachis hypogaea* L.) (syn.: *A. africana* Lour., *A. americana* Tenore, *Arachnida hypogaea* Moench)
- Quinoa, white quinoa (*Chenopodium quinoa* Willd.)

Less important and researched alternative crops are follows:

- Safflower, fals saffran (*Carthamus tinctorius* L.) (syn.: *Carduus tinctorius* Falk
- Fonio, fonio roise (Digitaria *exilis* Stapf) (syn.: *D. iburua* Stapf—Black)
- Finger millet, african millet, coracan millet, ragi (*Eleusine coracana*/ L./Gaertn.) (syn.: *Cynosurus coracanus* L., *E. stricta* Roxb.)
- Echinochloa: a) barnyard grass, (wild) millet, prickly grass (*Echinochloa* [*Panicum*] *crus-galli* [L.] P.Beauv.) and b) Japanese barnyard millet, sham millet, jungle rice (*E.* [*Panicum*] *colona* [L.] Link var. *frumentacea* Blatter et Mc Cann)
- Little millet, small millet (*Panicum sumatrense* Roth & Roemer & Schultes)
- Kodo millet, dith millet, kodo kodra (*Paspalum scrobiculatum* Lam.) (syn.: *P. kora* Willd., *Panicum dissectum* L.)
- Pearl millet, spiked millet, bulrush millet, cat's tail millet, bajri (*Pennisetum glaucum* L.) (syn.: *P. typhoideum* [Burm. f.] Rich., *Panicum glaucum* L.)
- Intermediate wheatgrass (*Elytrigia intermedia* [Host] Nevski) (syn.: *Thinopyrum intermedium* [Host] Barkworth & D. Dewey, *Agropyron i.* [Host] Beauv., *Elymus hispidus* [Opiz] Melderis)

- Fenugreek (*Trigonella foenum-graecum* L.) (syn.: *Foenumgraecum sativum* Medic.)
- Wild rice, Indian rice, Canadian rice (*Zizania palustris* L., *Z. aquatica* L.)
- Maybe some *Vigna* sp.

Throughout history, humankind has been directed toward discovering the new and unknown. Sometimes unknown crops or uncommon use present challenges for new generations. In Europe, four groups of development programs for using "new" plants have been utilized:

- Agricultural and industrial program financed by EU that dealt with biological characteristics of plants, production, industrial treatment, and processing. The following plants were included in these programs: *Cuphea* spp., *Simmondsia chinensis* (Link) Schneid., *Crambe abyssinica* Hochst. & R.E. Fries, *Limnathes* spp., coriander (*Coriandrum sativum* L.), *Dimorphotheca pluvialis* (L.) Moench, *Euphorbia lagascae* Spreng., rapeseed (*Brassica napus* L.), sunflower (*Helianthus annuus* L.), flax (*Linum* spp.), sorghum (*Sorghum* spp.), Jerusalem artichoke (*Helianthus tuberosus* L.), lupin (*Lupinus* spp.).
- Bilateral programs for exchange of information and heritage.
- National programs with collaborations of universities and institutes. German programs included study of the following plants: coriander (*Coriandrum sativum* L.), spurges (*Euphorbia* spp), rapeseed (*Brassica napus* L. with varieties*), chicory (*Cichorium intybus* L.), and sorghum (*Sorghum* spp.). A vast program in Great Britain included quinoa (*Chenopodium quinoa* Willd.). France, Spain, Italy, Greece, and Portugal studied plants such as hibiscus (*Hibiscus* spp.), castor bean (*Ricinus communis* L.), *Simmondsia chinensis* (Link) Schneid.) and *Cuphea* spp.
- Industrial programs for production and processing of aromatic, pharmaceutical, and bioactive plants [57].

Much attention has lately been dedicated to sorghum and millet for human consumption with detailed description, nutritional value, and production of all sorghum varieties (*Sorghum* spp.); among millet, the following varieties are described: pearl millet (*Pennisetum glaucum*), finger millet (*Eleusine coracana* L.), Italian millet (*Panicum italicum* L.), Indian millet (*Paspalum scrobiculatum* L.), broomcorn millet (*Panicum miliaceum* L.), little millet (*Panicum sumatrense* Roth & Roemer & Schultes), barnyardgrass (*Echinochloa crus-galli* [L.] P.B.), and jungle rice (*Echinochloa colona* [L.] Link) [58].

Austria studied Jerusalem artichoke (*Helianthus tuberosus* L.), hemp (*Cannabis sativa* L.), and *Crambe abyssinica* Hochst. & R.E. Fries [59]. In recent years, producers have been taught about numerous crop production niches. On the Austrian list in 1997 were the following plants: *Amaranthus caudatus* L. and other amaranth species, *Panicum miliaceum* L., *Papaver somniferum* L., *Carthamus tinctorius* L., *Linum usitatissimum* L., and *Cannabis sativa* L.

Sown as plant mixtures and intercropped plants, rare crops represent effective management strategies for the intensified conservation of rare crop species and the increase of biodiversity. However, the use of different options in organic farming depends on environmental and socioeconomic conditions, as well as locally available genotypes of rare crops. The preservation and development of new cultivars of rare crops is possible only within a framework of target-oriented genetic banks and breeding programs.

## References

1. http://europa.eu.int/comm/agriculture/qual/organic/index_en.html (accessed September 2005).
2. http://www.usda.gov/nass (accessed September 2005).
3. Le Guillou, G. and Sharpé, A., *Organic Farming*, Office for Official Publications of the European Commission, Luxembourg, 2001.
4. Guidelines for the Production, Processing, Labelling, and Marketing of Organically Produced Foods, Codex Alimentarius Commission, CAC/GL 32.1999, Point 7.
5. http://www.fao.org/arganicag/doc/Organic_perspectives.html (accessed October 2005).
6. Rigby, D. and Cáceres, D., Organic farming and the sustainability of agricultural systems, *Agric. Syst.*, 68 (1), 21, 2001.
7. http://ifoam.org/Principles_Organic_Agriculture.html (accessed October 2005).
8. Vogl, C.R., Kilcher, L., and Schmidt, H., Are standards and regulations of organic farming moving away from small farmers knowledge? *J. Sustain. Agric.*, 26 (1), 5, 2005.
9. Lampkin, N., Organic farming: Sustainable agriculture in practice, in *The Economics of Organic Farming: An International Perspective*, Lampkin, N. and Padel, S., Eds., CABI, Oxford, 1994, 468.
10. Torjusen, H. et al., Food system orientation and quality perception among consumers and producers of organic food in Hedmark County, Norway, *Food Qual. Prefer.*, 12 (3), 207, 2001.
11. Stockdale, E.A. et al., Agronomic and environmental implications of organic farming systems, *Advances in Agronomy*, 70, 261, 2001.
12. Bavec, M., *Ekološko kmetijstvo/Organic Farming* (Knjižnica za pospeševanje kmetijstva), Kmečki glas, Ljubljana, 2001.
13. http://www.ifoam.org/organic_facts/farming/index.html (accessed September 2005).
14. Bavec, M., Ekološka živla — tržna priložnost v Sloveniji (Organic foods — market possibility in Slovenia), in *Varnost živil*, presented at 22nd Food Technology Days 2004, dedicated to F. Bitenc, March 18–19, 2004, Radenci, Gašperlin, L., Žlender, B., Eds., Ljubljana, Biotehniška fakulteta, Oddelek za živilstvo, 2004, 153.
15. Willer, H. and Minou, Y., *The World of Organic Agriculture: Statistics and Emerging Trends*, International Federation of Organic Agriculture Movements (IFOAM), 7th ed., Bonn, Germany, 2005.
16. http://www.ams.usda.gov/nop/NOP/standards/ProdHandPre.html (accessed September 2005).

17. Hamm, U. and Gronefeld, F., Market situation for organic livestock products in Europe, in *Socio-Economic Aspects of Animal Health and Food Safety in Organic Farming Systems, Proceedings of the 1st SAFO Workshop,* September 5–7, 2003, Florence, Italy, M., Hovi, Martini, A., and Padel, S., Eds., 2003, 27.

18. Gambelli, D., Naspetti, S., and Vairo, D., Why are consumers buying organic meat and milk? A qualitative study of the Italian market, in *Socio-Economic Aspects of Animal Health and Food Safety in Organic Farming Systems, Proceedings of the 1st SAFO Workshop,* September 5–7, 2003, Florence, Italy, Hovi, M., Martini, A., and Padel, S., Eds., 2003, 140.

19. Thiele, S. and Weiss, C., Consumer demand for food diversity: Evidence for Germany, *Food Policy,* 28, 99, 2003.

20. Vaarst, M. et al., Sustaining animal health and food safety in European organic livestock farming, *Livest. Prod. Sci.,* 94 (1–2), 61, 2005.

21. Woese, K. et al., A comparison of organically and conventionally grown foods—results of review of the relevant literature, *J. Sci. Food Agric.,* 74, 281, 1997.

22. Worthington, V., Effect of agricultural methods on nutritional quality: A comparison of organic with conventional crops, *Alternative Therapies,* 4 (1), 58, 1998.

23. Alföldi, T., Bickel, R., and Weibl, F., Vergleichende Qualitätsuntersuchungen. Neue Ansätze und Impulse täten gut, Ökologie & Landbau, 117, 11, 2001.

24. Velimirov, A. and Müller, W., Die Qualität biologisch erzeugter Lebensmittel. Ergebnisse einer umfasenden Literaturrecherche zur Ermittlung potenzieller Vorteile Biologisch ezeugter Lebensmittel, Report Bio Austria, Ludwig Boltzmann Institut für Biologischen Landbau und Angewandte Ökologie, Wien, 2003.

25. Plochberger, K., Feeding experiments: A criterion for quality estimation of biologically and conventionally produced foods, *Agric., Ecosyst. Environ.,* 27, 419, 1989.

26. Velimirov, A., Kienzl-Plochberger, K. and Schwaiger, E., Futterwahlversuche mit Ratten und mikrobiologische Untersuchungen als integrative Testmethoden zur Ermittlung der Qualität landwirtschaftlicher Produkte, Förderdienst des Bundesministerium für Land- und Forstwirtschaft, Umweltschutz und Wasserwirtschaft, Wien, 2000.

27. Velimirov, A. et al., The influence of biologically and conventionally cultivated food on the fertility of rats, *Biol. Agric. Hortic.,* 8, 325, 1992.

28. Velimirov, A., Integrative methods of product quality assessment in connection with the p-value-determination, presented at Symposium in Lednice, Chechia, May 2002, 4.

29. Velimirov, A., *Biogramm und Qualitätskennzahl. Beiträge zur 7. Wissenschaftstagung zum Ökologischen Landbau in Wien,* Freyer, B., Ed., 2003, 233.

30. Loczisky, T., Vergleich von Lebensmitteln aus konventionellem und biologischem Landbau, *Ernte für das Leben,* 22, 1998.

31. Strutzmann, I., Biolandbau –Umweltschutz und Gesundheit Lebensmittel, Bio und Gesundhei, presented at 6th Alpe-Adria Biosymposium, Krastowitz–Klagenfurt, Austria, November 20, 2002, 9.

32. Rembialkowska, E., The impact of organic agriculture on food quality, *Agricultura,* 3 (1), 19, 2004.

33. news@dontjustsurvive.com (accessed November 2005).

34. Altieri, M.A., The ecological role of biodiversity in agroecosystems, *Agric., Ecosyst. Environ.*, 74 (1–3), 19, 1999.
35. Swift, M.J. and Anderson, J.M., Biodiversity and ecosystem function in agroecosystems, in *Biodiversity and Ecosystem Function*, Schultze, E., and Mooney, H.A., Eds., Springer, New York, 1993, 57.
36. Hole, D.G. et al., Does organic farming benefit biodiversity? *Biol. Conserv.*, 122 (1), 113, 2005.
37. Ivančič, A. et al., Agriculture in the Slovenian transitional economy: The preservation of genetic diversity of plants and ethical consequences. *J. Agric. Environ. Ethics.*, 16 (4), 337, 2003.
38. http://www.soilassociation.org/web/sa/saweb.nsf (accessed August 2005).
39. Booij, C.J.H. and Noorlander, J., Farming systems and insect predators, *Agric., Ecosyst. Environ.*, 40 (1–4), 125, 1992.
40. http://europa.eu.int/comm/agriculture/qual/organic/def/index_en.html (accessed October 2005).
41. Pacini, C. et al., Evaluation of sustainability of organic, integrated, and conventional farming systems: A farm and field-scale analysis, *Agric., Ecosyst. Eviron.*, 95 (1), 273, 2003.
42. Gosling, P. and Shepherd, M., Long-term changes in soil fertility in organic arable farming systems in England, with particular reference to phosphorus and potassium, *Agric., Ecosyst. Environ.*, 105 (1–2), 425, 2005.
43. http://www.ers.usda.gov./data/organic/ (accesed September 2005).
44. Caporali, F. and Onnis, A., Validity of rotation as an effective agroecological principle for a sustainable agriculture, *Agric., Ecosyst. Environ.*, 41 (2), 101, 1992.
45. Huxham, S.K., Sparkes, D.L., and Wilson, P., The effect of conversion strategy on the yield of the first organic crop, *Agric. Ecosyst. Environ.*, 106 (4), 345, 2005.
46. Schmidt, H. et al., Legume breaks in stockless organic farming rotations: Nitrogen accumulation and influence on the following crop, *Biol. Agric. & Hortic.*, 17 (2), 159, 1999.
47. Knudsen, M.T., Comparison of interspecific competition and N use in pea-barley, faba bean-barley, and lupin-barley intercrops grown at two temperate locations, *J. Agric. Sci.*, 142, 617, 2004.
48. Bavec, F. et al., Competitive ability of maize in mixture with climbing bean in organic farming, in *Researching Sustainable Systems*, presented at First Scientific Conference of the International Society of Organic Agriculture Research (ISOFAR), Adelaide, Sept. 21–23, 2005; Köpke, U. et al., Eds., IFOAM, Adelaide, Australia, 2005, 502.
49. Radics, L. and Pusztai, P., Shading effect as a weed management strategy in crop rotations, *Novenytermeles*, 49 (1–2), 69, 2000.
50. Spadaro, D. and Gullino, M.L., Improving the efficacy of biocontrol agents against soilborne pathogens, *Crop. Prot.*, 24 (7), 601, 2005.
51. Fowler, C. and Mooney, P., *Shattering: Food, Politics and the Loss of Genetic Diversity*, University of Arizona Press, Tucson, AZ, 1990.
52. Francis, C.A., *Multiple Cropping Systems*, MacMillan, New York, 1986.
53. Gliessman, S.R., Sustainable agriculture: An agroecological perspective, *Adv. Plant Pathol.*, 11, 45, 1995.
54. Williams, J.T., *Cereals and Pseudocereals*, Williams, J.T., Ed., Chapman and Hall, London, 1995.

55. Janick, J. et al., Diversifying U.S. crop production, in *Progress in New Crops*, Janick, J., Ed., ASHS Press, Alexandria, VA, 1996, 98.
56. Bavec, F. *Nekatere zapostavljene in /ali nove poljščine (Some of Disregarded and/or New Field Crops*), Univerza v Mariboru, Fakulteta za kmetijstvo, Maribor, 2000.
57. Van Soest, L.J., New crop development in Europe, in *New Crops*, Janick, J. and Simon, J.E., Ed., Wiley, New York, 1993, 30.
58. FAO, *Sorghum and Millets in Human Nutrition*, Rome, 1995.
59. BMLF–Bundensministerium für Land und Forstwirtschaft. *Forschungsbericht 1996*, Abteilung II a 1, Wien. 1997, 23.

*chapter two*

# Cereals

The most important alternative cereals for organic production are spelt (*Triticum spelta* L.), einkorn (*Triticum monococcum* L.), emmer (*Triticum dicoccum* [Schrank] Schübl.), and KAMUT® (see description in Chapter 4.1), (Figure 2.1).

## 2.1  Spelt

### 2.1.1  Introduction

Spelt (*Triticum spelta*) is also called large spelt, dinkel, dinkel wheat, and German wheat.

Two controversial hypotheses exist regarding spelt origins. According to the first hypothesis, spelt can be traced back to the area now known as Iran. The second hypothesis traces spelt's origins to geographically separate areas: Iran and northeastern Europe. Another hypothesis supports the idea

*Figure 2.1* Spikes of spelt, einkorn, emmer, and kamut®, in order from left.

that spelt developed from cultivated tetraploid emmer wheat (*Triticum turgidum* L. var. *dicoccum* [Körn.] Bowden) with AABB genomes and from wild wheat (*Triticum tauschii* [Coss.] Schmal. Ag. Squarosa) with DD genomes [1].

During the centuries when profitable cereals such as common wheat (*Triticum aestivum* L. ssp. *vulgare*) and durum wheat (*Triticum durum* L.) were not grown in large quantities, spelt was extremely important for peoples' survival. At the end of the eighteenth century, spelt was produced in the areas close to primary centers of cereal origins in Asia (Turkey, Iran, Iraq, Pakistan, and Afghanistan) and in most European countries. Spelt production in European countries decreased over time. Reproduction began in the last couple of decades of the twentieth century in hilly and mountainous regions of central European countries, in amounts that allowed for the preservation of indigenous species and the breeding of new ones. In Switzerland, spelt was written about approximately 100 years ago and was produced last in Murtemberg [2]. In Slovenia, it disappeared after World War II.

Because it is an undemanding and adjustable crop, spelt is the most appropriate cereal for organic production. Due to increasing consumer awareness, spelt is becoming a very important bread cereal anywhere in the world where controlled organic farming exists. Merchants, traders, and producers from Central America and Canada have also expressed great interest.

## 2.1.2   Botany

### 2.1.2.1   Systematic and genotypes

Spelt (*Triticum spelta* L., syn. *T. aestivum* [L.] Thell. Subsp. *spelta* [Thell.] Mk.; *T. vulgare* Vill. *spelta* Alef., *T. sativum spelta* Hackel, *T. zea* Host, *Spelta vulgaris* Seringe, *T. arduini* Mazz.) belongs to the family of *Poaceae* (grasses), genus *Triticum* L., and the group of hulled wheats [3]. The seeds and hulls of hulled wheats remain together after harvesting. Emmer (*T. dicoccum* [Schrank] Schübl) and einkorn (*T. monococcum* L.) also belong to the group of hulled wheats. Hulled wheats are included in the Italian subgroup of farro wheats ("farro" is an ethnobotanical Italian name). Farro wheats include *T. sinskajae* A. Filat. et Kurk, but this subspecies classified by Szabó and Hammer [4], or species described by Miller [5], is naked seeded. Spelt is hexaploid (2n = 6x = 42) wheat with genomic structure AABBDD.

Few spelt varieties are grown as land race populations. The first spelt variety suggested for growth in Europe was Bauländer Speltz, whereas Schwabenkorn, Altgold, Ostro, Frankenkorn, Oberkulmer, and Holstenkorn are often grown in German-speaking countries. The most common varieties in Belgium are Hercule, Redonte, and Roquin and, in the Netherlands, Gotro. The standard Swiss varieties are Ostro, Oberkulmer, and Lueg, and the newest ones are Balmegg, Ostar, and Setel [6]. Compared with standard varieties, Hubel achieves a 25% higher yield.

Spelt wheat is more resistant to diseases and infections than other varieties of common wheat. However, the disease resistance of spelt depends on genotypes. For example, the Hubel variety sets a very strong standard for plant infections with fungus (*Puccinia striiformis, P. recondita,* and *eptorisa nodorum*), but its resistance of powdery mildew (*Erysiphe graminis*) is similar to that of the varieties Ostro and Oberkulmer. Research conducted on the resistance of 20 varieties to powdery mildew *Erysiphe graminis* and *Puccinia recondita* showed that at the stages of small plats (EUCARPIA: EC 10–20, Zadoks 10–20 [7]) Bauländer Speltz was the only variety resistant to the *E. graminis,* while none of the varieties were resistant to the *P. recondita.* Cultivars Altgold Rotkorn and Rouquin showed fewer infection symptoms than Ostro and Oberkulmer Rotkor. Bauländer Speltz and Ostro both lack the dominant genes needed for resistance to *P. recondita* [8].

### 2.1.2.2   Morphology

Spelt is a tiller plant with many secondary stems, which are hollow and range from 1.4 to 1.7 meters tall. Spelt leaves are longer, narrower, less hairy, or have no hair at all in comparison with common wheat. The rachis of the spelt seed head is fragile, a condition that enables its spreading. Spike is similar to a grass weed *Agropyron repens* L.; spikelets are set in opposite directions. Spikes are medium brittle to brittle, and they are thinner and longer than common wheat. Spikelets contains up to six flowers that most often produce two seeds after self-pollination. These seeds are surrounded by hulls after harvesting and are called hulled seeds. Glumes are attached to the seeds so firmly that they are not removed after harvest and, consequently, the rachis breaks. Awn is attached to glumes in some cultivars. Glumes that surround the spikelet are transversally cut and have a short and obtuse outgrowth in the shape of a tooth in the middle. Glumes may account for as much as 25 to 45% of the spikelet's entire mass. Dehulled spelt seed is compressed from the side, round on the back side, and has a significantly well-expressed beard (Figure 2.2).

## 2.1.3   Growth and ecology

The growth and development of spelt are marked in a usual way (Zadoks, EC stages). Spelt is cold resistant and can survive lower temperatures than other common wheat cultivars, and it is less sensitive to long-term snow. Production is possible at 1200 meters above sea level. Spelt is an adjustable plant and can be grown on all soil textures; therefore, it is suitable for extensive production. The appropriate soil pH is between 6 and 7.

## 2.1.4   Organic cultivation practice

Due to relatively successful disease resistance, spelt is an appropriate crop for organic farming, though it is important to take into consideration the eventual surplus of nitrogen in soil due to previous legumes or potential excessive mineralization. In this case, spelt usually lodges. Spelt can be placed after all other crops in crop rotation, including cereals, and can be sown as a less-demanding plant in the second or third year; after that, at

*Figure 2.2* Spelt grains.

least a three-year pause is recommended. It can be sown after corn, beets, lettuce, and even buckwheat. Classic soil preparation with plow and presowing tillage, similar to the preparation for wheat, is suitable. Dry, uncompressed, and lumpy soil is not recommended, despite the adjustability of spelt and its successful emergence even after rough presowing tillage. In cases of regulated water-air soil ratio, plowing is not necessary and direct seeding is possible. We have to be careful regarding the nutrient content, however, and keep in mind that demands for phosphorus and potassium — as well as other nutrients — are similar to those of wheat. Uptake depends on the amount of yield. When we are making decisions regarding presowing nitrogen fertilization (compost included), it is necessary to know nitrogen content and to at least conduct the nitrate soil test. Where more than 20 mg kg$^{-1}$ soil $NO_3$-N is present, nitrogen fertilizers are not added.

Spelt is sown as a winter cereal in Europe, whereas it is sown as a summer cereal in America and most of Canada. In temperate climates, spelt can be sown later in the season than common wheat. Spelt germinated in Slovenia in the Zreško Pohorje hills when sown in the late autumn, just before snowfall. Nevertheless, late sowing dates should be an exception and not a rule. If we take into consideration a general rule regarding the appropriate development of cereals for hibernation, we have to sow spelt early enough for it to reach the stage of three true leaves (EC 23) before winter frost. Judging from a height above sea level, we can conclude that winter spelt in a continental climate should be sown at least 2 months before permanent snow arrives. There are no comparable results regarding optimal sowing dates where more winter and spring sowing dates are checked, but only such a comparison could provide proper justification for winter sowing. The advised seeding rate is 400 seeds m$^{-2}$ or 160 to 250 kg of dehulled seed ha$^{-1}$ in some cases. In different production circumstances within the eastern U.S., 90 to 112 kg of hulled seed ha$^{-1}$ are recommended, and there are no differences noted between seeding rates of 67 to 100 kg seeds ha$^{-1}$ without irrigation and 100 to 134 kg seeds ha$^{-1}$ with irrigation [6, cited in 1].

Swiss results show that seeding rates of 150, 200, and 250 kg of sown spelt seeds ha$^{-1}$ present only a small difference in yield [9]. Codianni et al. [10] investigated spelt at two seeding rates (200 and 300 viable seeds per square meter) and found higher yield (12–28%) with the highest seeding rate. Similarly, Castagna et al. [11, 12] reported no significant effect of seeding rates between 200 and 400 seeds m$^2$ for einkorn, emmer, and spelt grown in north Italy (Sant'Angelo Lodigiano location, Lombardia region), while a significant increase of yields (10–16%) with the highest seeding rate was noted at two locations situated in central Italy. Spelt grain yield and spike number increased as seeding rate increased. Consequently, the best performance at the highest seeding rate resulting in a hulled grain yield of 3.09 t ha$^1$ and 479 spikes m$^{-2}$. These results agree with results described by Codianni [11]. However, in works [12–14], spelt sown at 300 seeds m$^2$ showed lower performance than that achieved in the present experimentation, signifying that 200 seeds m$^2$ seems to be the best crop condition for these wheats. In the case of Troccoli and Codianni [15], spelt showed the best yield when it was sown at 200 seed m$^2$, but it is important to consider that a range of factors may contribute to hulled wheats not achieving their potential yield. These include the seeds' germination, climate, and soil moisture conditions, as well as the presence of insects, disease, and weeds. In addition, plant characteristics, such as height, heading, tillage, and lodging resistance, combined with other traits (sink–source relationship, water use efficiency, etc.), may be important factors that diversify the performance of these species. With spelt possessing a two-to-three-times more productive tillering coefficient, we can state that in optimal growth conditions, a recommended 400 seed m$^{-2}$ for one-half to one-third lower plant density and equally lower seeding rates are advised. However, we [16] conclude that an equal yield of dehulled grains is achieved by different sowing rates (200, 400, 600 seeds m$^{-2}$), but the grain yield with hulls was significantly higher at the sowing rate of 400 seed m$^{-2}$. This means that for seeds without hulls, the higher 1000 seeds weight sown by 200 seeds m$^{-2}$ is the main reason for yield compensation and higher quality of yield at low plant population.

Spelt is sown at a similar depth as wheat but must not go below 4 cm (6 cm in light sandy soil), especially if sowing occurs late. Seed should also not be sown shallowly or left on the surface. As with all cereals, some seeds will be decayed by frost, and some remaining seeds will not grow productively and will develop secondary roots. Plants that are sown late, along with smaller plant density, provide space for weed competitiveness. Hulled or unhulled seeds may be sown, and growers will probably choose the same interrow spacing as with wheat (12 cm). However, because of thinner sowing spaces, up to 22 cm are given. Growers should make the decision regarding interrow spacing based on the planned number of germinated seeds m$^{-2}$ required for them to achieve as square a grain arrangement as possible.

With regard to crop provision, spelt should be harrowed if the soil is crusted immediately after sowing. Harrows should be used at EC 13 growth stage but not during emergence. In organic farming, the use of harrows at EC 20 to EC 35 growth stages is almost obligatory. Besides using harrows to control weeds in spelt, a lower amount of nitrogen fertilizers or compost should be added in organic farming, in case of foreseen modest mineralization. Special attention should be paid to the usage of quick nitrate tests, as is done with common wheat. In tests, the Ostro cultivar increased yield while increasing the rates from 0, 40, 60, 80, and 100 kg N ha$^{-1}$ [9]; it is important to provide enough available nitrogen from organic compounds in the soil.

## 2.1.5   Harvesting

Spelt is harvested in the middle of summer, before full maturity. We should not wait for spelt to reach full maturity or beyond, as with common wheat. Darker spelt straw is a sign that the spelt is overripe, and the yield will suffer a severe loss at harvesting. The harvesting should be done at night and on humid mornings, and soft layers should be used on the combine to prevent a rough impact on the stems or spikes of lodged plants.

In organic farming, we can expect 2 to 3 t ha$^{-1}$ of hulled grain yield with a harvest index (hulled grain: biological yield of plant mass) of approximately 0.52. Older literature sometimes states that it is not possible to store spelt the same way as wheat, because it is prone to spoiling and pests. Spelt should be storied in glumes and dried well. Hulled seed should contain less than 15% moisture.

## 2.1.6   Chemical composition, nutritional, and health value

In general, the proximate composition of spelt grain compared with common winter wheat, triticale, shows that spelt has a significantly higher content of protein. At other sites, the proximate composition and nutritional value of spelt depend on genotypes and environmental circumstances (see Table 2.1), and sometimes the values of spelt are similar or lower than those of conventional wheats [17]. According to Moudry et al. [18], the average content of N compounds in spelt grain was about 0.5% higher than in common wheat grain, and fiber content was 0.35% higher. Significant differences exist in fiber and fat content among spelt cultivars (Table 2.1). The fat content in the grains varied from 0.8 [19] to 4.7% in the case of Marconi et al. [20], where the total fiber varied from 10.5 to 14.9% dry base. Spelt cultivars' grain also contained a high proportion of monounsaturated fatty acids in the total fatty acid composition, with an average of 21.5% in contrast to 12.1% in common wheat and 13.7% in triticale [21].

*Table 2.1* Proximate Composition of Spring [22] and Winter Spelt (% at 10% Moisture in the Seeds) [17, 20]

| Component | Abdel-Aal et al. [22] | Ranhotra et al. [17][a] | Marconi et al. [20][a] |
|---|---|---|---|
| Number of varieties | 5 (spring) | 15 | 5 |
| Protein | 13.9 (13.4–14.4) | 16.7 (14.2–22.9) | 14.2 (12.9–16.6) |
| Carbohydrates | 59.3 (57.1–60.8) | 60.4 (53.3–62.8) | 58.7 (53.5–62.5) |
| Fat | 2.0 (1.19–2.23) | 1.5 (0.8–2.2) | 4.0 (3.4–4.7) |
| Fiber | 9.0 (8.8–9.3) | 9.5 (9.3–10.5) | 11.4 (9.4–13.4) |
| Ash | 1.6 (1.51–1.66) | 1.8 (1.7–2.2) | 1.71 (1.43–1.89) |

[a] Adapted by Cubadda and Marconi [23] with permission.

*Source:* From Belton, P. and Taylor, J., *Pseudocereals and Less Common Cereals: Grain Properties and Potential*, 2002. Copyright 2005, with kind permission of Springer Science and Business Media.

*Source:* Reprinted from *Cereal Chem.*, 72 (6), Abdel-Aal, E.S.M., Hucl, P., and Sosulski, F.W., Compositional and nutritional characteristics of spring einkorn and spelt wheats, 621, 1995, Copyright 2006, with permission from American Association of Cereal Chemists.

*Source:* Reprinted from Ranhotra, G.S., Gelroth, J.A., and Lorenz, K.J., *J. of Food Composition and Anal.*, 9, Copyright (2006), with permission from Elsevier.

*Source:* Reprinted from *Cereal Chem.*, 76 (1), Marconi, E. et al., Kernel properties and pastamaking quality of five European spelt wheat (*Triticum spelta* L.) cultivars, 25, 1999, Copyright 2006, with permission from American Association of Cereal Chemists.

Genetic influence on differences in protein content of spelt dehulled grains compared to common winter wheat shows that spelt has a significantly higher content of crude protein, total proteins, and more nonessential amino acids, especially proline and glutamic acid content (16 g N, $P < 0.05$). The assessed values show that with an increasing content of crude protein, the content of proteins does not increase proportionally [24].

Amino acid composition of spelt is similar to the composition of common wheats (Table 2.2). In both, the most limiting amino acid is lysine. The content of lysine varies among spelt cultivars, which ranges from 2.35 [25] to 3.96 [17]. The high lysine content of some spelt products with respect to the same products manufactured with common wheats can be attributed to the lower degree of refinement of spelt flour, since spelt is manly used as whole or pearled grain meal [24].

Four grain samples of spelt cultivars grain contained 427 mg P kg$^{-1}$ DM, and the levels of some micro (trace) elements — especially Cu, Mn, Zn, and Cu — were higher than in common wheat (Table 2.3) [21]. Accessions [22, 28, 29] in other spelt were found to have higher phosphorus levels than common wheat and emmer.

With regard to vitamin composition in spelt, the authors [22, 28] found similar conclusions: some B-vitamin (thiamine, niacin) values were higher in spelt than in common wheat. The data for the content of vitamin E (tocopherols) are different; in the case of Grela [21], vitamin E activity is 143% higher than in common wheat, but Marconi [30] found that the content of tocopherols in spelt was lower than in other wheats.

*Table* 2.2 Essential Amino Acid Composition of Spelt (g 16 g N$^{-1}$) According to Different Authors

| Amino Acid | Grela et al. [21] | Jorgensen et al. [26] | Bonafacia et al. [27] | Range for Wheat [21, 26, 27] |
|---|---|---|---|---|
| Histidine | 2.33 | 2.43 | 2.20 | 2.30–2.70 |
| Isoleucine | 4.16 | 3.90 | 3.60 | 3.30–3.90 |
| Leucine | 7.06 | 6.69 | 6.60 | 6.6–7.15 |
| Lysine | 3.19 | 2.73 | 2.60 | 2.84–3.02 |
| Methionine | 1.70 | 1.65 | 1.70 | 1.62–1.70 |
| Phenylalanine | 4.51 | 4.62 | 4.90 | 4.5–5.05 |
| Tyrosine | 3.53 | 2.91 | 2.30 | 2.68–3.31 |
| Threonine | 3.70 | 2.82 | 2.60 | 2.90–3.46 |
| Tryptophane | 1.49 | – | – | 1.42 |
| Valine | 5.02 | 4.64 | 4.40 | 4.35–4.76 |

*Source:* Reprinted from *J. Sci. Food Agric.*, 71 (3), Grela, E.R., Nutrient composition and content of antinutritional factors in spelt (*Triticum spelta* L.) cultivars, 399, 1996. Copyright Society of Chemical Industry. Reproduced with permission. Permission is granted by John Wiley & Sons Ltd. On behalf of the SCI.

*Source*: Reprinted from Bonafacia, G. et al., *Food Chem.*, 68, 437, 2000. Copyright (2005), with permission from Elsevier.

*Table* 2.3 Mineral Content in Spelt and Common Wheat

| Mineral | Ronhotra et al. [19] Spelt | Wheat | Grela [21] Spelt | Wheat |
|---|---|---|---|---|
| Phosphorus | 462 | 388 | 427 | 334 |
| Calcium | 29 | 37 | 43 | 55 |
| Potassium | 457 | 383 | 493 | 463 |
| Magnesium | 148 | 140 | 147 | 116 |
| Manganese | 4.8 | – | 5.12 | 3.6 |
| Copper | 0.4 | 0.5 | 1.10 | 0.45 |
| Selene | – | – | 0.37 | 0.43 |
| Zinc | 4.1 | 2.2 | 4.72 | 2.58 |

mg 100 g$^{-1}$ Dry Matter

*Source*: Reprinted from Ranhotra, G.S. et al., *Cereal Chemistry*, 73, 533. Copyright (2005), with permission from American Association of Cereal Chemists.

*Source:* Reprinted from *J. Sci. Food Agric.*, 71 (3), Grela, E.R., Nutrient composition and content of antinutritional factors in spelt (*Triticum spelta* L.) cultivars, 399, 1996. Copyright Society of Chemical Industry. Reproduced with permission. Permission is granted by John Wiley & Sons Ltd. On behalf of the SCI.

The average content of fat (lipid) in spelt differs, according to the authors [17, 20–22, 29], and reaches 4.4% on dry base, but in most cases it is still significantly higher than in common wheat. The fatty acid composition of spelt varieties is uniform, as reported by Grela [21] and Marconi [30]. The structure of fatty acids is as follows: about 55% of linoleic (18:2), 20.4% of oleic (18:1), 18.5% of palmitic (16:0), and 3% of linolenic (18:3) fatty acid [21]. The portion of oleic acid is significantly higher in spelt than in wheat (11.3%) and triticale (12.9%), but the portion of linoleic and linolenic acids is higher in common wheat and triticale.

Spelt is not a gluten-free food, but it is said to have a beneficial effect on the immune system due to mucopolysaccharides. Abdel-Aal et al. [31] conclude that spelt is not a safe grain for people with celiac disease, contrary to the implications of labeling bread made from spelt as "an alternative to wheat." All the different classes of gliadins — alpha, beta, gamma, and omega — appear to be active in celiac disease; therefore, it is reasonably certain that the spelt gliadin is also toxic. However, others claim that spelt is safe for individuals suffering from celiac disease, especially when food from whole-grain flour is used.

## 2.1.7 Processing

The procedure used to hull the seeds is important: glumes are more attached with autochthon cultivars than with new, often-crossed cultivars. Usually we hull with machines consisting of two vertical rotating stones, which are separated enough to remove glumes without crushing the seeds. Hulled seeds are further cleaned with stationary threshers and grain separators. Modern methods constitute centrifugal huskers, grinding on emery cones, grinding in snail transmission through emery rotor, and friction grinding with small turns and high pressure on seeds [32].

"Grünkern" is a traditional specialty made from green seeds harvested at the stage of milk maturity when the seeds contain about 40% moisture. After the harvest, the seeds must be dried at a temperature between110 and 160°C. This provides a significant beech wood aroma of and allows the seeds to preserve their greenish color. This type of seed is used as a special additive when making bread and soups, and a coffee substitute made from green spelt seeds is another delicious tradition.

Spelt can be used in the same way as other cereals, and the processing is also similar. The milling performance as determined by yield, damaged starch, ash, and particle-size distribution in the flour is uniform among the cultivars [20].

From a nutritional point of view, grinding spelt with millstones and using freshly ground flour is recommended. We mill the flour from whole seeds more often than with common wheat and only rarely use white flour from endosperm. In this case, the average flour yield is about 70%. When grinding into white flour, vitamin and fat content are reduced and only one-fifth of the ashes are left. Spelt flour contains less gluten of lower quality;

the gluten content is similar to that of other wheat [22]. Gluten properties, which are assessed by gluten index and sedimentation, as well as by alveograph and pharinograph parameters, vary widely among cultivars [33]. This is connected to the presence of common wheat genes, a consequence of new age breeding. Due to a lower glutinous quality, the baking properties of spelt flour are less suitable than common wheat flour. Rheological and baking properties of spelt flour are also influenced by the spelt genotype.

Sensory and chemical evaluations of the pastas, however, indicate that spelt is suitable for obtaining good-quality pasta [20]. Pasta cooking quality depends on protein quantity and quality. The spelt quality of proteins is a missing factor, but not the main one, because the quality of pasta also depends on the processing details. For example, the cooking quality of spelt pastas dried at a temperature of 90°C show good values of organoleptic judgments and total dry matter in comparison to drying at 60°C [25]. The combination of a high protein content and the high-temperature drying cycle adopted in pasta production could be responsible for these good results [20].

Flour mixture made from common wheat or gluten-rich spelt should increase the quality of pasta. Using a cross between wheat and spelt, the population mean of the dough parameters shifted toward the spelt parent [34].

Spelt bread represents an important organic specialty, to which white bread scoring standards, crumb grain, and texture cannot be compared. Baking quality standards — including the content of wet gluten, its extensibility and swelling, the content of N-substances, the sedimentation test, the falling number, and the content of starch — differ among cultivars. Cultivars strongly influence the sedimentation value, which can vary from 15 to 25 [6]. The data described by Bojnanska and Francakova [35] shows that Schwabenkorn and Rouquin are considered the most suitable highest-quality cultivars. There were good baking quality results for Baulander Spelz, as well. Unexpectedly bad results have been found with Rouquin, which showes the lowest water-absorbing capacity of flour, along with the lowest bread volume, specific volume, and baking extraction; the bread's tastewas only acceptable. The baking that resulted from Holstenkorn was evaluated as excellent. Conversely, Abdel et al. [36] reported that soft spelt flours have a short mixing time and potential applications in two-layer flat breads. Hard spelt flours have low gluten strength, but when produced like flaked cereal, they have an appealing flavor and texture.

Spelt bread quality depends on dough-making and baking procedures. The dough from spelt flour becomes soft and sticky after kneading and firmer over time. The baking procedure can start very early after the bread rises; however, it is often believed that the dough needs a long resting time, i.e., a long fermentation at moderate temperatures.

Besides using flour as all other real cereal flour types (pasta "Spätzle," flat cakes, pasta), spelt can be also used to make flakes, as malt in beer production, as a coffee substitute, or as roasted or dried "Grünkern" seeds

*Figure 2.3* Bread made from whole grained flour spelt.

at milk maturity stage. The latter is a very popular bread and soup supplement among consumers of organically produced food.

Organic flakes with optimal crisp and an easy breaking texture for breakfast, without any additional ingredients, are possible to produce [36]; different kinds of breakfast flakes (sweet, salted, paprika flavor, cocoa flavor, with amaranths) were produced with a blend of maize starch, grits, malt, salt, sugar, and different proportions of wholemeal or white spelt flour.

A modern organic diet suggests using whole grains. It is possible to use 100% whole grain flour for bread making (Figure 2.3), or an appropriate portion of whole grain flour should be added to the bread prepared from white flour (spelt or mixtures of other cereals). The spelt whole grain flour or whole grains should be soaked in water one day prior to making the bread. Spelt grain may also be used as a side dish or as an adequate substitute for rice, salad (Figure 2.4), or even cake with cream and honey. In this case, the pericarp should be removed by a grinding machine.

## 2.2 Einkorn

### 2.2.1 Introduction

*Einkorn* (wheat) is a German name for one-grained wheat, also called small spelt (*Triticum monococcum*, syn: *T. monococcum* susp. *monococcum* [3, 37], *T. vulgare bidens* Alef., *niviera monococcum* Ser.) [38].

Einkorn originated in the Near East, primarily in the Euphrates and Tigris cereal regions. The wild spelts (*T. beoticum*, *T. aegilopoides* — one-grained wheats; *T. thaoudar* and *T. urartu* — two-grained wheats) and ancestors of *T. monococcum*, which has the genetic constitution AA, were harvested in the Paleolithic Age, 16000–15000 BC [1]; the collecting of these seeds continued into the early twentieth century. Historically, einkorn was cultivated in the Caucasus Mountains, in the Mediterranean region, and in

*Figure 2.4* Salad made from spelt grains.

northwestern Europe. Einkorn was also the first grown cereal in the Balkan region. Einkorn was an important cultivated cereal in temperate environments and marginal agricultural lands through mideastern and southwestern Europe. Around 1970, einkorn was grown in only a few isolated regions of Europe, but einkorn later became the subject of intensive breading and marketing programs, initially in the U.S. [38]. Today, einkorn production (Figure 2.5) is limited to small regions within Turkey, India, and the U.S., and to small organic farms through the world — mainly in Italy, Austria, Swiss, Germany, and Slovenia.

*Figure 2.5* Packages of einkorn and spelt products.

## 2.2.2   Botany

The average einkorn plant height is about 1 meter, depending mostly on moisture supply. Under dry conditions, the plant may reach a lower size. Cultivated einkorn contains a single seed in spikelets. Glumes are long and narrow and fall away from the hulled seed. During harvesting, spikes fall into hulled spikelets. The dehulled grains are generally small (21.2–37.4 mg seed$^{-1}$) and have a very soft endosperm texture (hardness indexes –7.3 to 27.2) [39].

## 2.2.3   Cultivation practices

An increase in the seeding rate (100, 150, and 200 viable seeds m$^{-2}$) of einkorn progressively decreased its grain yield (1.69, 1.45, and 1.13 t ha$^1$, respectively). In contrast, however, the spike number increased. It appears that einkorn requires more spacing among plants to give a good grain yield, while significant yield losses may occur as a consequence of competition among plants when high seed density is used. For this reason, the best performance of einkorn is obtained with a seeding rate of 100 seeds and 360 spikes per square meter, respectively [15]. This result disagrees with Codianni et al. [10], who found einkorn to have, under the same climate conditions, the highest grain yield with high plant density (300 and 400 seeds m$^{-2}$).

The plant needs of nitrogen are very low: conventional fertilization trials showed yield reduction when the nitrogen fertilizer rates increased, or yield was unaffected by different nitrogen rates [10, 11].

Under adverse Italian growing conditions, einkorn selections produced protein with a yield equal to or higher than barley and durum wheat [40, 41]. Other works reported that einkorn has lower grain yield compared to durum wheat grown in northern and southern Italy [11, 13]. The grain yields are low, but in the last decade of the twentieth century, they varied between 120 to 4200 kg grain yield ha$^{-1}$.

## 2.2.4   Chemical composition and processing

The protein content of einkorn ranges from 10 to 26%; as a rule, this value is higher than in barley, durum wheat, and common hard red wheats. Amino acid composition is similar to common wheats [15]. According to Loje et al. [39*], the chemical composition and functional properties of 10 einkorn cultivars show high ash content (2.3–2.8% dry matter), variable protein content (10.3–19.5% dry matter), and low content of (13;14)-ß-glucan (0.29–0.71% dry matter). The content of total dietary fiber was low (7.6–9.9% dry matter) compared to common wheat. The content of lysine in selected samples varied considerably (1.51–3.15 g lysine 100 g$^{-1}$ protein). Abdel Aal et al. [22, 36]

---

* Reprinted from *J. Cereal Sci.*, 37 (2), Løje, H. et al., Chemical composition, functional properties and sensory profiling of einkorn (*Triticum monococcum* L.), 231, Copyright 2005, with permission from Elsevier.

reported that ancient wheats were comparable to modern wheats in chemical composition, except for einkorn protein content.

The high protein einkorn accession was similar to common hard red spring wheat in kernel size, but soft grain gave low flour yields. Einkorn flour had low sedimentation values, weak Mixograph curves, and low loaf volumes [31]. Most of the einkorn samples had high falling numbers (mean 362 sec) and high amylograph viscosities (mean 1185 BU) [39]. Despite high protein content, einkorn flour showed very weak gluten strength and potential applications in soft-wheat products in which yellow color is not a problem [36]. However, biomass yield, grain protein, and SDS sedimentation values of the hulled wheats were significantly influenced by nitrogen, which had no effect on grain yield [42].

Cooked einkorn seeds are softer in consistency and less mealy, adhesive, and fibrous than samples of spelt and wheat. Due to its high gelatinization viscosity, high protein content, low dietary fiber content, good consistency, and pleasant flavor, cooked grains may be used as a substitute for rice or as a weaning food [38].

## 2.3   Emmer

### 2.3.1   Origin, botany, and history

Two-grained hulled wheat, also called emmer in German languages, originated in the Near East [43]. Emmer (*Triticum dicoccum*) was named using numerous synonyms, such as *T. dicoccon* Shuebl., *T. turgidum* (L.) Thell. ssp. *dicoccum* (Schrank) Schübl., *T. farrum* Bayle-Barrele, *T. amyleum* Seringe, *T. zea* Wagini, *Spelta amylea* Seringe, *T. volgense* (Flaskb.) Nevski, *T. vulgare dicoccum* Alef., *T. sativum dicoccum* Hack., and incl. *T. ispahanicum* Helsot) [3]. Two-grained wheats represent tetraploid and hexaploid species. The spike of the two-grained wheat is predominantly made up of spikelets, each consisting of two well-developed seeds. Long and narrow glumes with sharp beaks are characteristic of emmer. The rachis of the emmer is fragile, and spikelets readily disarticulate when harvested at full maturity [44].

Extensive research has shown that remnants of wild emmer in early civilization sites date from the late Paleolithic Age, 17000 BC [45]; from the late Mesolithic and early Neolithic Ages (Stone Age), 10000 BC; and from the Bronze Age, 10000 to 1000 BC [45–53], found mainly in the Far and Near East, Northern Africa, and Europe. Emmer production was preserved in isolated regions in Ethiopia, India, and south central Russia [47, 54].

### 2.3.2   Production

Troccoli and Codianni [15] indicate that emmer is the most appropriate hulled wheat species for production in marginal area preservation or when the cultivation of economically profitable crops is precluded by water deficiency and soil poorness.

Experimental field trials have shown that emmer has a lower grain yield than durum wheat but higher than einkorn and spelt, also under high infestation of *Avena fatua* and *Phalaris arundinacea* [13, 14]. Codianni et al. [10] found that yield was reduced for emmer as nitrogen fertilizer rates increased, or it was unaffected [11]; this is an important fact for low input production systems such as organic farming. Codianni et al. [10] compared two seeding rates (200 and 300 viable seeds m$^{-2}$) and found that emmer produced better grain yield (12–28%) with the higher seeding rate. Similarly, Castagna et al. [11] reported no significant effect of seeding rates between 200 and 400 seeds m$^2$ for emmer in northern Italy, while a significant increase of yields (10–16%) was produced with the highest seeding rate in two locations in central Italy. The data [1, 10, 11, 14] showed that 200 seeds m$^2$ seems to be the optimal crop condition for this wheat. Troccoli and Codianni [15] also reported that emmer was superior in terms of grain yield and kernel weight, and demonstrated earlier heading when sown at 200 seed m$^2$.

Emmer grain yields vary the most between 200 and 4000 kg ha$^{-1}$, and a firm correlation exists between grain yield and climatic circumstances (location). An estimated grain yield of emmer may achieve between 45 to 75% of the grain yield of spring wheat [23].

### 2.3.3 Utilization

The protein content of emmer seeds is 5 to 30% higher than that of oats and barley seeds [1]. Emmer is not a gluten-free cereal.

Before learning that emmer flour produces a good loaf of bread, early civilizations consumed emmer as a porridge. Breads produced from whole grain flour are heavilyy textured with a milder tastethan breads made from triticale or rye flour. The loaf and texture of emmer bread is not equal to that of breads made from common wheat [23]. However, this kind of organic product has a special taste and a high content of nutritional compounds.

## 2.4 Kamut

KAMUT®* is a registered trademark used for marketing products that are made with this remarkable grain. The Kamut grain's ability to produce high quality without artificial fertilizers and pesticides makes it an excellent crop for organic farming [55], although its history and taxonomy may be disputed. Kamut is described as *Triticum turgidum* ssp. *turanicum*, also called Khorasan wheat, but some taxonomists suggest the names *T. turgidum* ssp. *polonicum* and Egyptian durum wheat (*T. turgidum* ssp. *durum*). Wheats of this type originate from Egypt to the Tigris-Euphrates valley; the Kamut Association of North America (KANA) promotes the legend that 36 giant grain kernels were originally taken from an Egyptian tomb. A Montana wheat farmer then

---

* Registered trade mark of Kamut International, Ltd.

*Figure 2.6* Kamut® grains.

planted and harvested a small crop and displayed the grain as a novelty at the local fair [55*, 1**].

Kamut has a large (Figure 2.6), high-quality grain with high gluten in protein content. The grains are used for all products, like common wheat, as well as couscous and bulgur, due to the similar kernel characteristics to durum wheats.

## 2.5  Triticale

### 2.5.1  Introduction

Plant breeders call triticale a bread cereal of the future. Triticale (×*Triticosecale* Wittm. & Camus; sin. ×*T. rimpaui* Wittm., ×*Triticale* Müntzing) is an artificial cereal species created from a cross between two genus, wheat (*Triticum* sp.) and rye (*Secale cereale* L.). Triticale is appropriate for production in areas that are less favorable for wheat production; it is a highly promising cereal in hilly regions where wheat and barley production presents a certain risk. It can also successfully replace extensive rye cultivars. The main advantages of triticale are its resistance to low temperatures and the possibility of high grain yield [56].

* Reprinted from Kamut: Ancient grain, new cereal, in *Perspectives on New Crops and New Uses*, Quinn, R.M., 182; with permission from ASHS.
** Reprinted from Alternative wheat cereals as food grains: einkorn, emmer, spelt, kamut and triticale, in *Progress in New Crops*, Stallknecht, G.F., Gilbertson, K.M., and Ranney, J.E., 156, with permission from ASHS.

## 2.5.2 History, taxonomy, and cultivars

The first wheat and rye crosses date back to 1875 and 1876, when the first crosses resulted in sterile hybrids. The first fertile hybrid with reproductive capabilities was reported in Germany between 1888 and 1891. Triticale production began in Russia in 1918, and the term "triticale" was first used in Germany in 1935. Between 1935 and 1936, a discovery was made that all forms were amphiploids and all crosses were octoploids (2n = 42), not diploids (2n = 14). Germ culture technique was first used in crossing soon afterward, in 1940. After 1950, intensive selection began in Europe (in Slovenia F. Jesenko) and in North America [57, 58].

Triticale is a product of hybridization based on A- and B-genomes of tetraploid wheats or A-, B- and D-genomes combinations of hexaploid wheats with rye genome –R. Fertile offspring is possible with amphiploids (AABBRR or AABBDDRR) from genetic points of view, which appeared with duplication of the chromosome number. Hexaploid triticale synthesized from wheats (AABB) and rye (RR) are called primary hexaploids, while hexaploid triticale synthesized from crosses of hexaploid triticale and/or hexaploid wheats or octoploid triticale are called secondary hexaploid triticale. In the process of crossing different wheat and rye species, the following triticale crosses were synthesized: *Triticale aestivum* Shulind-octoploid triticale (common wheat and rye hybrid, 2n = 56), *Triticale durum* Shulind-hexaploid triticale (hard wheat and rye hybrid, 2n = 42), and *Triticale trispecies* Shulind-hexaploid triticale (hybrids with properties of hard wheat, common wheat, and rye, 2n = 42).

More recently, the tetraploid triticale has been synthesized (2n = 28, AARR). The new term ∞*Triticosecale* is used for hexaploid triticale (2n = 42, AABBRR) [56].

Triticale is characterized by height and, consequently, modern selection is directed into the inclusion of semi-recessive genes for the reduction of stem height $Rht_1$ in $Rht_3$ of *Triticum aestivum*. These genes can reduce stem height up to 40%. For the same purpose, dominant genes of rye EM 1 are used.

The following cultivars have been established throughout the world: Mexican Armandillo, Canadian Rosner, American 6 TA, Hungarian N° 57 and N° 64, and Soviet cultivars AD 206, AD 201, AD 196, and AD 1. Clercal winter cultivar is recommended in Slovenia, but French Triman has taken its place in the European market. During official Slovenian testing in 1995, Clercal showed better qualities than the cultivar Almo. Sandro from Switzerland, which was included in Austrian species list, is one of the more interesting early cultivars. In Poland, the following cultivars, which are resistant to low temperatures, have been chosen: Prego, Ugo, Tewo and Moreno, LAD 285, B86-3398, and octoploid Lanca x medium height rye [59]. Cultivars that are resistant to aluminum, such as Tahara, Tahara S, and Abacus, were also chosen [60]. The new triticale cultivars have improved agronomic traits including high yields, resistance to lodging and ergot

(*Claviceps purpurea* [Fr.] Tul.), and plump kernels. The new cultivars Modus, Dato (from Germany), Presto (from Poland), and SV 92280 (from Sweden) are especially productive, resistant to leaf and stem rust and powdery mildew, and produce good grain quality [61].

## 2.5.3   Morphology

Plant habit is of intermediary type and does not return to original components. Wheat plants will only occasionally appear with some cultivars. In triticale, wheat genomes prevail (octoploids 3:1, hexaploids 2:1); this is why triticale is similar to wheat if the properties of spike, its color, and color of grain are considered. Octoploid triticale has either winter or spring common wheat habit (*Triticum aestivum*), while hexaploid triticale is similar to durum wheat (*Triticum durum*). Both vary significantly in plant height, tend to tiller less, and generally have larger spikes than wheat. Triticale also displays rye properties: violet colored coleoptiles, stem brittleness below the spike, elongated glumes, an increased number of spikelets in spikes, and an often increased stem height. Triticale glumes are wider than rye glumes, and there are more grains in triticale spikes than rye spikes. Stem height, spike length, and spike compaction are, in most cases, intermediary properties. Studies show that 30% of triticale properties are not inherited from parent species, including the brittleness of the triticale spike head at maturation and grain deformation. The normal hectoliter weight of first class quality is 58 to 72 kg hl$^{-1}$, but cultivars with up to 78 kg hl$^{-1}$ are also known.

## 2.5.4   Production and yielding

Triticale was introduced into 75 countries in 1975, and field sizes covered with this crop increased from 100,000 to 500,000 hectares. In 1986, over one million hectares of field were covered, and by 1991–92, triticale was produced in 31 countries on 2.4 million hectares (8). The highest quantities are produced in Poland, Russia, Germany, the U.S., France, Australia, Bulgaria, and South Africa [62].

## 2.5.5   Growth and ecology

Triticale adapts more readily to low temperatures and dry conditions than other cereals. Giunta et al. [63, 64] report on the lack of significant differences between the irrigated and rainfed treatments; their findings support the strong adaptation abilities of triticale to water deficits when compared to to wheat under similar environmental conditions.

Triticale is sown as winter cereal in Russia, East Asia, and European countries; in Mexico, North America, East Africa, India, and Australia, it is sown as spring cereal. Winter triticale regenerates and grows faster in spring than winter wheat. Variations in thermal rate with a base temperature ($T_b$) of 0°C and duration of spikelet and floret initiation and spikelet number differ between triticale and wheat, but no such differences have been observed for maximum or final floret number. Significant positive relationships between the thermal rate of spikelet and floret initiation and photoperiod have been found for triticale and were consistent for all tillers. The duration of spikelet and floret initiation varied more than the rate of initiation among years, sowing dates, and tillers, particularly in triticale [65*]. Winter triticale required 11% more thermal time from sowing to anthesis and had a 25% lower maximum grain weight than that of spring genotypes. The maximum grain filling rate was 40% higher in spring than in winter triticale [66].

Longer grain-filling periods are noted in triticale. Grain-filling duration was mainly controlled by environmental conditions and was not correlated with grain yield [66].

For some cultivars, such as Lasko, an extremely short vernalization period is noted; some cultivars do not need the vernalization period at all. Breeders strive to create photoperiodically neutral plants that can be grown regardless of geographical position and season. In connection with draught resistance, triticale is also resistant to high temperatures. Cells successfully retain water, and plants have a very useful wax surface and well-developed root system. Triticale is more demanding than rye regarding water supply but gives normal yield in conditions typical for rye production.

Two critical periods appear during triticale growth and development: a period of sexual cell development (gametogenesis) and a period of embryo and endosperm formation (embryogenesis–endosperm genesis). Moisture deficit can cause disturbance with gamete formation and fullness of grain in spikes 6 to 10 days before spike (head) emergence in meiosis. Triticale is most sensitive in this period. Grain deformation has also occurred as a consequence of different factors, such as increased α amylase activity and poor chromosome regulation in the early stages of endosperm development. Deformed grains are often half empty or the endosperm filling is extremely poor. Grain germination in spikes is another characteristic of triticale, and subjection to germination is conditioned by a high degree of α amylase in grain.

Physiological leave senescence is slower with triticale than with other cereals. Consequently, 1 to 2% fewer assimilates in wheat grains and 3 to 5% fewer assimilates in rye grains are synthesized than in triticale grains over the same time period [56].

## 2.5.6 Organic cultivation practice

### 2.5.6.1 Crop rotation
Triticale crop depends on crop rotation to a great extent. Russian results confirm this statement; the highest yields were achieved after the following

* Reprinted from Field Crops Res., 47 (2-3), Ewert, F., Spikelet and floret initiation on tillers of winter triticale and winter wheat in different years and sowing dates, 155, Copyright 2005, with permission from Elsevier.

precrops: pea, early potato cultivars (probably also late ones), grass-clover mixtures, and grainy legumes. Following the winter wheat precrops on humus, a reduction of spreading, spike size, and perfection was noticeable. Crops that followed this precrop were covered with weeds. Extreme yield reduction was also noted when sowing triticale after corn. Thus, triticale presumably requires soil preparation at least two weeks before sowing; late sowing after corn can reduce yield by one-half [56].

### 2.5.6.2  Soil

Triticale produces especially good results on light, sandy soil or on soil with acid reaction. Literature on the subject states that triticale's demands regarding soil are lower than those of wheat, but this applies to less fruitful octoploid triticale; hexaploid triticale's demands are similar to those of wheat. Some recently created cultivars are resistant to salty soil, and cultivars that are extremely adjustable to production in acid or basic soil have also been synthesized. Triticale can adapt more easily to acid soil than wheat; its toxic tolerance regarding aluminum and iron is also higher [67].

Soil that is too dry at sowing can result in slower development, poor spreading, and lower yield. Basic and precrop cultivation is the same for triticale as for wheat, with special importance placed on timely presowing preparation.

### 2.5.6.3  Sowing and plant cultivation practice

In most cases, grain does not preserve good germination properties; therefore, it has to be stored at low temperatures and in low relative air humidity. Sowing depth should be 4 to 6 cm and density should range from 500 to 600 viable seeds $m^{-2}$ in optimal humidity conditions. The approximate amount for sowing is 180 kg of seed $ha^{-1}$. While the literature says that in different growing conditions, seeding rates from 80 kg of seed $ha^{-1}$ are used, some authors state that the optimal number of well-developed plants at the end of winter is between 200 and 250. By all means, seed quantity depends on cultivar spreading abilities, sowing date, and soil conditions. Triticale is sown at the same time as rye 500 meters above sea level; in wheat areas and on well-fertilized soil, triticale should not be sown in early autumn. Early sowing can result in conditions that are too favorable for the growth and development of plants in fall. Before winter, 3 to 6 side tillers should be formed.

According to Giunta and Motzo [64*], increasing triticale sowing rates in spring (50, 100, 300, 500, and 700 viable seeds $m^{-2}$) caused faster development, with a 1-week difference between extreme sowing rates in time to achieve double ridge, terminal spikelet, and increased grain yield from 524 to 781 g $m^2$ on average by increasing the sowing rate from 50 to 300 seeds $m^{-2}$. Sink size was also responsible for the higher harvest index observed at

---

* Reprinted from *Field Crops Res.*, 87 (2-3), Giunta, F. and Motzo, R., Sowing rate and cultivar affect total biomass and grain yield of spring triticale (*Triticosecale* Wittmack) grown in a Mediterranean-type environment, 179, Copyright 2005, with permission from Elsevier.

the highest population densities in two seasons out of three. Harvest index is affected during the postanthesis period — depending on plant-population density — through its effects on sink capacity, which is the function of both the number of kernels and the number of stems per unit area produced after anthesis.

According to our experiences, triticale reacted well to fertilization with stable manure. Due to a lack of research on triticale fertilization in organic farming, certain general knowledge must be considered. Nutrient demands of triticale have been said to fall somewhere between those of wheat and rye. Triticale grows extremely fast in spring; therefore, it needs more nitrogen than wheat. General comparison of nutrients, e.g., nitrogen fertilizers, have shown that triticale uses more nitrogen than other crops, including wheat [68], but Kaerpenstein-Machan et al. [69] report that triticale is advantageous in low-input systems. The data from Ewert and Hohermaier [70] supports the conclusion that spikelet initiation and number per ear are most stable in triticale treatment without any additional nitrogen application. However, their results also indicate that nitrogen application in triticale can increase the spikelet number.

Weed management can be provided by one or two mechanical treatments done by weed comb. Triticale has expressed resistance to cereal pathogens and needs no special care, especially if appropriate crop rotation is taken into consideration.

Harvesting has one phase, which is carried out with cereal combines during the transition from wax to full maturity.

### 2.5.7 Nutritional and health value

Stallknecht [1] asserts that the nutritional quality of triticale is superior to wheat. The higher ash content, lower milling yields of flour, inferior loaf volume, and texture distract from the commercial baking use of triticale. Triticale flour is rich in proteins (average of 13–15%) and high levels of the amino acid lysine, suggesting a promising use for the production of human foods.

### 2.5.8 Processing

Kernel quality is a complex combination of physical and chemical characteristics whose expression depends on their genetic nature and the influence of environment [71, 72].

Triticale is often included in prepared mixed-grain hot and cold cereals, muffin flours, and cracker products. For bread making, flour from triticale is less suitable due to the lower quality of glutinous, an elastic protein complex of glutens made of gliadins and glutenins. Transformation of starch into sugar in triticale flour is very fast due to high alpha amylase activity, which results in a sweet bread. Cultivars selection requires a mixture of triticale and wheat flour in ratios of 50:50, 20:80, or 30:70. Bread composition from these types of flour is suitable for use. Pena and Amaya [73] indicate that triticale flour blends of up to 50% with wheat flours produce breads of similar quality to those made from wheat flours only. Baking tests have shown that many cultivars of hexaploid triticale can be used in bread making, by mixing the wheat flour with up to 70% triticale flour [61].

Some authors and practitioners make bread solely from triticale flour. Considering the indices characterizing the bread-making properties (falling number, protein content, glutenin subunit composition, Zeleny number, water absorption capacity, and bread volume) of many triticale cultivars, the cultivars Moreno, Presto, Tewo, Dato, and SV 92280 provide promising results when using its flour for bread making [61].

## 2.6 Intermediate wheatgrass

Intermediate wheatgrass, also called Wild triga (*Thinopyrum intermedium* [Host] Barkworth & D. Dewey), is a native grass that spread from southern Europe to the Middle East, southern parts of Russia, and western Pakistan. During Byzantine times, this grass was utilized as a grain crop in Turkey, Armenia, and part of Russia [74]. During recent decades of intermediate wheatgrass development in the U.S., it has become a more interesting and unique perennial grain crop for human consumption. In addition to the rich nutritional value of the grains, their ecological benefits (low input crop, perennial crop effectively prevents soil erosion) are very important. Wheatgrass is relatively resistant to disease and is very suitable for organic production, in spite of the fact that intermediate wheatgrass can be affected by foot rot caused by *Fusarium graminearum* or *Bipolaris sorokiniana*. However, progressive improvement in seedling survival was made in three populations selected for their resistance using three cycles of recurrent phenotypic selection [75].

The plant is a perennial grass "wheatgrass" (fam. *Poaceae*), first described as *Agropyron intermedium* (Host) Beavois. On the basis of cytogenetic information and biological relationships, wheatgrass was classified as *Thinopyrum intermedium* Barkworth & D. Dewey. Synonyms of previous classifications are *Elytriga intermedia* (Host) Nevski, *Agropyron intermedium* (Host) Beauv, and *Elymus hispidus* (Opiz) Malderis [76]. Many refined cultivars already exist in this species, such as Luna, Oahe, Chief, Greenar, Tegmar, Greenleaf, Topar, Manska, Reliant, Haymaker, and Beefmaker.

Intermediate wheatgrass is a rhizomatous grass, growing 1.0 to 1.5 m high, with pubescent and glabrous forms [77]. Grain weight varies from 3.5 to 9.0 mg. The golden-brown colored seeds average from 5 to 6 mm in length and 1 to 2 mm in width.

Intermediate wheatgrass is rich in crude proteins [77]. The cultivar Oahe content comprises 20.8% of proteins, 3.21% of fats, and 2.64% of ash. The intermediate wheatgrass protein is nutritionally limiting in lysine, the same as wheat, but has higher levels of all other essential amino acids than wheat, with the absence of significant amounts of antinutrients. In this case no gluten was found [78], but low contents of gluten were sometimes detected.

This pseudocereal grows best in dry areas with only 200 or 400 mm of annual precipitation. The optimal sowing period is after harvesting cereals, while warm-season weeds are not competitive at that time of year. Wheatgrass can be rotated with other annual crops on a 5 to 7 year basis, and the first harvest will come approximately 1 year after sowing. The sowing rate is only 15 to 20 kg seeds ha⁻¹. Wheatgrass is sown at interrow space from 15 to 20 cm, or from 60 to 1000 cm by using 7 kg seeds ha⁻¹. Wheatgrass emerges in 3 to 4 days, assuming sufficient soil moisture. It ripens in the middle of summer and takes 4 to 7 days to achieve uniform ripeness. The specific suggestion for the organic production of intermediate grass is to use stable manure instead of slurry and liquid manure for fertilization, in order to avoid bad soil structure and dock weed (*Rumex* sp.). Especially on farms without animal fertilizers, plant mixtures with white clover due to nitrogen fixation are essential.

Combining should be done when the stalks and spikes are completely dry. After harvest, hay can be fed to horses and cattle that do not demand voluminous fodder. At harvest, it can remain 100 to 600 or 800 kg grain ha⁻¹.

Stone milling resulted in flour with farinograph characteristics more similar to those of whole wheat flour than did impact and roller milling. Endosperm recovery rates of intermediate wheatgrass are lower (40–50%) than in wheat (70–75%) due to smaller seed size [78].

The seeds can be cooked like wild or brown rice to make a flavorful and attractive cooked-grain product. Baking products made from wheatgrass are sweet, nutty, and of high quality. Bread, muffins, cookies, and cakes have favorable appearances, textures, flavors, and overall characteristics [78, 79].

## *References*

1.  Stallknecht, G.F., Gilbertson, K.M., and Ranney, J.E., Alternative wheat cereals as food grains: Einkorn, emmer, spelt, kamut and triticale, in *Progress in New Crops*, Janick, J., Ed., ASHS Press, Alexandria, VA, 1996, 156.
2.  Nowacki, A., *Getreidebau*, Verlagsbuchhandlung Paul Parey, Berlin, 1905, 244.
3.  Ivančič, A. *Hibridizacija pomembnejših rastlinskih vrst (Hybridization of Important Plant Species)*, Univerza v Mariboru, Fakulteta za kmetijstvo, Maribor, 2002.
4.  Szabó, A.T. and Hammer, K., Notes on the taxanomy of farro: *Triticum monococcum, T. dicoccum* and *T. spelta*, in *Hulled Wheats*, Proc. of the 1st Int. Workshop on Hulled Wheats, Castelvecchio, Tuscany, Italy, July 21–22, 1995, Padusoli, S., Hammer, K., and Heller, J., Eds., IPGRI Rome and IPK Gatersleben, 1996, 2.
5.  Miller, T.E., Systematic and evolution, in *Wheat Breeding — Its Scientific Basis*, Lupton, F.G.H., Ed., Chapman and Hall, London, 1987, 1.
6.  Bavec, F., Spelt (*Triticum spelta* L.), in *Nekatere Zapostavljene in/ali Nove Poljščine (Some of disregarded and/or new field crops)*, Univerza v Mariboru, Fakulteta za kmetijstvo, Maribor, 2000, 94.

7. Zadoks, J.C., Chang, T.T., and Konzak, C.F., Decimal code for growth stages of cereals, *Weed Res.*, 14 (6), 415, 1974.

8. Zeller, F.J. et al., Untersuchungen zur Resistenz des Dinkel-Weizens (*Triticum aestivum* [L.] Thell. ssp. *spelta* [L.] Thell.) gegenüber Mehltau (*Erysiphe graminis* f. spp. *tritici*) und Braunrost (*Puccinia recondita* f. spp. *tritici*), *Die Bodenkultur*, 45 (2), 147, 1994.

9. Maillard, A., Techniques culturales et productivite de l'epeautre en Suisse romande, *Revue-Suisse-d'Agriculture*, 26 (2), 77, 1994.

10. Codianni, P. et al., Agronomical performance of farro in southern Italy environments (in Italian), *Inf. Agrar.*, **38**, 45, 1993.

11. Castagna, R. et al., Performance of farro under different growing conditions (in Italian), *Inf. Agrar.*, **35**, 52, 1993.

12. Castagna, R. et al., The farro: Nitrogen fertilization and seeding rate assessments (in Italian), *Inf. Agrar.*, **35**, 44, 1994.

13. Codianni, P. et al., Performance of selected strains of "farro" (*Triticum monococcum* L., *Triticum dicoccon* Schübler, *Triticum spelta* L.) and durum wheat (*Triticum durum* Desf cv. Trinakria) in the difficult flat environment of southern Italy, *J. Agron. Crop Sci.*, 176, 15, 1996.

14. Troccoli, A. et al., Agronomical performance among farro species and durum wheat in a drought-flat land environment of southern Italy, *J. Agron. Crop Sci.*, 178, 211, 1997.

15. Troccoli, A. and Codianni, P., Appropriate seeding rate for einkorn, emmer, and spelt grown under rainfed condition in southern Italy, *Eur. J. Agron.*, 22 (3), 293, 2005.

16. Rantaša, J., Performance of the pelt (*Triticum aestivum* L. ssp. *spelta* MacKey) yield depending on sowing rate of unshelled and shelled seed and sowing date (Abstract), thesis, University of Maribor, Faculty of Agriculture, Maribor, 2004.

17. Ranhotra, G.S., Gelroth, J.A., and Lorenz, K.J., Nutritional composition of spelt wheat, *J. Food Comp. Anal.*, 9, 81, 1996.

18. Moudry, J. and Dvoracek, V., Chemical composition of grain of different spelt (*Triticum spelta* L.) varieties, *Rostl. Vyroba.*, 45 (12), 533, 1999.

19. Ranhotra, G.S. et al., Nutritional profile of three spelt wheat cultivars grown at five different locations, *Cereal. Chem.*, 73, 533, 1996.

20. Marconi, E. et. al, Kernel properties and pasta-making quality of five European spelt wheat (*Triticum spelta* L.) cultivars, *Cereal Chem.*, 76 (1), 25, 1999.

21. Grela, E.R., Nutrient composition and content of antinutritional factors in spelt (*Triticum spelta* L.) cultivars, *J. Sci. Food Agric.*, 71 (3), 399, 1996.

22. Abdel-Aal, E.S.M, Hucl, P., and Sosulski, F.W., Compositional and nutritional characteristics of spring einkorn and spelt wheats, *Cereal Chem.*, 72 (6), 621, 1995.

23. Cubadda, R. and Marconi, E., Spelt wheat, in *Pseudocereals and Less Common Cereals: Grain Properties and Utilization Potential*, Belton, P., and Taylor, J., Eds., Springer-Verlag, Berlin, 2002.

24. Chrenkova, M. et al., Assessment of nutritional value in spelt (*Triticum spelta* L.) and winter (*Triticum aestivum* L.) wheat by chemical and biological methods, *Czech J. Anim. Sci.*, 45 (3), 133, 2000.

25. Marconi, E. et al., Composizione amminoacidica di sfarinati integrali e raffinati di spelta (*Triticum spelta* L.), in *Materie prime transgeniche, sicurezza alimentare e controllo qualità nell'industria cerealicola*, Proceedings of the symposium, Campobaso, Oct. 7–8, 1999, Cubadda, R., and Marconi, E., Eds., Compabaso, 2000, 184.

26. Jorgensen, J.R., Olsen, C.C., and Christiansen, S., Cultivation and quality assessment of spelt (*Triticum spelta* L.) compared with winter wheat (*Triticum aestivum* L.), in *Small Grain Cereals and Pseudo-Cereals*, Stolen, O. et al., Eds., European commission, Luxemburg, 1997, 31.

27. Bonafacia, G. et al., Characteristics of spelt wheat products and nutritional value of spelt wheat-based bread, *Food Chem.*, 68, 437, 2000.

28. Ranhotra, G.S., et al., Baking and nutritional qualities of a spelt wheat sample, *Lebensm.-Wiss.+Technol.*, 28, 118, 1995.

29. Piergiovanni, A.R. et al., Mineral composition in hulled wheat grains: A comparison between emmer (*Triticum dicoccon* Schrank) and spelt (*T. spelta* L.) accessions, *Int. J. Food Sci. Nutr.*, 48, 381–386, 1997.

30. Marconi, E., et al., Qualitative and quantitative evaluation of lipidic fraction in *Triticum dicoccon* Schrank and *T. spelta.-L.*, *Tec. Molitoria*, 52 (8), 826, 2001.

31. Abdel-Aal, E.S.M. et al., Kernel, milling and baking properties of spring-type spelt and einkorn wheats, *J. Cereal Sci.*, 26 (3), 363, 1997.

32. Kaiser, H.G., Einsatz der Schmirgel- und Friktionsschleifmaschine bei der Dinkelschalung, *Getreide Mehl und Brot*, 10, 296, 1995.

33. Marconi, E. et al., Spelt (*Triticum spelta* L.) pasta quality: Combined effect of flour properties and drying conditions, *Cereal Chem.*, 79 (5), 634, 2002.

34. Zanetti, S. et al., Genetic analysis of bread-making quality in wheat and spelt, *Plant Breeding*, 120 (1), 13, 2001.

35. Bojnanska, T. and Francakova, H., The use of spelt wheat (*Triticum spelta* L.0 for baking applications, *Rostl. Vyrob.*, 48 (4), 141, 2002.

36. Abdel-Aal, E.S.M., Hucl, P., and Sosulski, F.W., Food uses for ancient wheats, *Cereal Foods World*, 43 (10), 763, 1998.

37. Szabó, A.T. and Hammer, K., Notes on the taxanomy of farro: *Triticum monococcum*, *T. dicoccum* and *T. spelta*, in *Hulled Wheats*, Proc. of the first Int. Workshop on Hulled Wheats, Castelvecchio, Tuscany, Italy, July 21–22, 1995, Padusoli, S., Hammer, K., and Heller, J., Eds., IPGRI Rome and IPK Gatersleben, 1996.

38. Bavec, F., Einkorn wheat (*Triticum monococcum* L.), in *Nekatere zapostavljene in /ali nove poljščine (Some of disregarded and/or new field crops)*, Univerza v Mariboru, Fakulteta za kmetijstvo, Maribor, 2000.

39. Løje, H. et al., Chemical composition, functional properties and sensory profiling of einkorn (*Triticum monococcum* L.), *J. Cereal Sci.*, 37 (2), 231, 2003.

40. Vallega, V., Field performance of varieties of *Triticum monococcum*, *T. durum* and *Hordeum vulgare* grown at two locations, *Genet. Agric.*, 33, 363, 1979.

41. http://www.farro.com (Accessed April 2006).

42. Castagna, R. et al., Nitrogen level and seeding rate effects on the performance of hulled wheats (*Triticum monococcum* L., *T. dicoccum* Schubler and *T. spelta* L. evaluated in contrasting agronomic environments, *J. Agron. Crop Sci.*, 176 (3), 173, 1996.

43. Nevo, E., Genetic resources of wild emmer wheat revisited: Genetic evolution, conservation and utilization, Proc. Int. Wheat Genetics 7th Symp., Cambridge, UK, August 11–13, 1988, 121.

44. Bavec, F., Emmer (*Triticum dicoccum* Schrank), in *Nekatere zapostavljene in/ali nove poljščine (Some of disregarded and/or new field crops)*, Univerza v Mariboru, Fakulteta za kmetijstvo, Maribor, 2000.
45. Zohary, D. and Hopf, M., Domestication of plants in the Old World, the origin and spread of cultivated plants in West Asia, Europe, and the Nile valley, 2nd ed., Oxford Univ. Press, New York, 1993.
46. Helmqvist, H., The oldest history of cultivated plants in Sweden, *Opera Bot.*, 1, 1, 1995.
47. Harlan, J.R., The early history of wheat: Earliest traces to the sack of Rome, in *Wheat Science Today and Tomorrow*, Evans, L.T. and Peacock, W.J., Eds., Cambridge Univ. Press, Cambridge, 1981.
48. Bottema, S. and Ottaway, B.S., Botanical, malacological and archaeological zonation of settlement deposits at Gomolava, *J. Archeol. Sci.*, 9 (3), 221, 1982.
49. Van der Veen, M., Botanical evidence for Garamantian agriculture in Fezzan, southern Libya, *Rev. Paleobotany Palynology*, 73 (1–4), 315, 1992.
50. Bakels, C.C., The botanical shadow of two early Neolithic settlements in Belgium: Carbonized seeds and disturbances in a pollen record, *Rev. Paleobotany Palynology*, 73 (1–4), 1, 1992.
51. Hopf, M., Plant remains from Bogazköy, Turkey, *Rev. Paleobotany Palynology*, 73 (1–4), 99, 1992.
52. Küster, H., Early Bronze Age plant remains from Freising, southern Bavaria, *Rev. Paleobotany Palynology*, 73 (1–4), 205, 1992.
53. Peña-Chocarro, L., et al., The oldest agriculture in northern Atlantic Spain: New evidence from El Mirón Cave (Ramales de la Victoria, Cantabria), *J. Arch. Sci.*, 32 (4), 579, 2005.
54. Perrino, P. and Hammer, K., *Triticum monococcum* L. and *T. dicoccum* Schubler (Syn of *T. dicoccon* Schrank) are still cultivated in Italy, *Genet. Agr.*, 36, 343, 1982.
55. Quinn, R.M., Kamut: Ancient grain, new cereal, in *Perspectives on New Crops and New Uses*, Janick, J., Ed., ASHS Press, Alexandria, VA, 1999, 182.
56. Bavec, F., Triticale (*Triticosecale* Wittm. & Camus), in *Nekatere zapostavljene in/ali nove poljščine (Some of disregarded and/or new field crops)*, Univerza v Mariboru, Fakulteta za kmetijstvo, Maribor, 2000.
57. Villareal, R.L., Varughese, G., and Abdalla, O.S., Advances in spring triticale breeding, *Plant Breed. Rev.*, 8, 43, 1990.
58. Hürlein, A.J. and Valentine, J., Triticale (*Triticosecale*), in *Underutilized Crops. Cereals and Pseudocereals*, Williams, J.T., Ed., Chapman & Hall, London, 1995, 274.
59. Banaszak, Z., Marciniak, K., and Brykczynska, L., Evaluation of frost resistance methods using tests breeding materials at DANKO. Biul. Instit. Hod. i Aklim. Roslin, Racot. No. 1998, 205.
60. Zhang, X.G. and Jessop, R.S., Analysis of genetic variability of aluminium tolerance response in triticale, *Euphythica*, 102 (2), 177, 1998.
61. Tohver, M. et al., Quality of triticale cultivars suitable for growing and bread-making in northern conditions, *Food Chem.*, 89 (1), 125, 2005.
62. http://faostat.fao.org/; http://www.nsac.ns.ca/pas/gradstudents/sve/Triticale %20History.html (accessed October 2005).
63. Giunta, F., Motzo, R., and Deidda, M., Effect of drought on yield and yield components of durum wheat and triticale in a Mediterranean environment, *Field Crops Res.*, **33**, 399, 1993.

64. Giunta, F. and Motzo, R., Sowing rate and cultivar affect total biomass and grain yield of spring triticale (×*Triticosecale* Wittmack) grown in a Mediterranean-type environment, *Field Crops Res.*, 87 (2–3), 179, 2004.

65. Ewert, F., Spikelet and floret initiation on tillers of winter triticale and winter wheat in different years and sowing dates, *Field Crops Res.*, 47 (2–3), 155, 1996.

66. Santiveri, F., Royo, C., and Romagosa, I., Patterns of grain filling of spring and winter hexaploid triticales, *Eur. J. Agron.*, 16 (3), 219, 2002.

67. Royo, C., Rodriquez, A., and Romagosa, I., Differential adaptation of complete and substituted triticale, *Plant Breeding*, 111, 113, 1993.

68. Sylvester-Bradley, R., Opportunities for lower nitrogen inputs without loss of yield or quality: An agronomic and economic appraisal, in *Proc. of the Cereals RetD Conf.*, Cambridge, H-GCA, London, 1993, 198.

69. Karpenstein-Machan, M., Honermeier, B., and Hartman, F., *Triticale. Production Aktuel*, DLG-Verlag, Frankfurt/Main, 1994.

70. Ewert, F. and Honermeier, B., Spikelet initiation of winter triticale and winter wheat in response to nitrogen fertilization, *Eur. J. Agron.*, 11 (2), 107, 1999.

71. Johansson, E. and Svensson, G., Variation in bread-making quality: Effects of weather parameters on protein concentration and quality in some Swedish wheat cultivars grown during the period 1975–1996, *J. Sci. Food Agric.*, **78**, 109, 1998.

72. Johansson, E., Effect of two wheat genotypes and Swedish environment on falling number, amylase activities, and protein concentration and composition, *Euphytica*, **126**, 143, 2002.

73. Pena, R.J. and Amaya, A. Milling and breadmaking properties of wheat-triticale blends. *J. Sci. Food Agr.*, 60, 483, 1992.

74. Bavec, F., Intermediate Wheatgrass, in *Nekatere zapostavljene in/ali nove poljščine (Some of Disregarded and/or new Field Crops)*, Univerza v Mariboru, Fakulteta za kmetijstvo, Maribor, 2000, 215.

75. Krupinsky, J.M. and Berdahl, J.D., Selecting resistance to *Bipolaris sorokiniana* and *Fusarium graminearum* in intermediate wheatgrass, *Plant Dis.*, 84 (12), 1299, 2000.

76. Wagoner, P., Intermediate wheatgrass (*Thynopyrum intermedium*), development of perennial grain crop, in *Underutilized Crops–Cereals and Pseudocereals*, Chapman and Hall, London, 1995, 247.

77. Karn, J.F. and Berdahl, J.D., Nutritional, morphological and agronomic characteristics of selected intermediate wheatgrass clones, *Can. J. Plant Sci.*, 64 (4), 909, 1984.

78. Becker, R. et al., Compositional, nutritional and functional evaluation of intermediate wheatgrass (*Thinopyru intermedium*), *J. Food Process. and Preservation*, 15 (1), 63, 1991.

79. Wagoner, P., Perennial Grain — new use for intermediate wheatgrass, *J. Soil Water Conserv.*, 45 (1), 81, 1990.

# chapter three

# Pseudocereals (without millets)

## 3.1 Buckwheat

### 3.1.1 Introduction

Common buckwheat (*Fagopyrum esculentum* Moench, syn. *F. sagitatum* Gilib., *F. vulgare* T. Nees, Hill) used to grow primarily in Asia (Pendzab, Tibet, and Poamur regions); today, however, wild plants can be found in China (Himalaya Mountains), Siberia, and the Far East. Chinese people planted buckwheat as early as the eleventh century BC, and the diploid genotype of buckwheat has been grown in Mongolia since the tenth century. Buckwheat was first mentioned in European countries in 1396, and it has been grown in the U.S. ever since the seventeenth century.

Three different explanations exist for the naming of buckwheat. The origins of the English word "buckwheat," the Netherlands term "boekweit," and the German term "Buchweizen" can be traced to the Greek name "Phagos" for beech combined with the word wheat, because the seed of buckwheat is similar to the tetrahedron form of beech seed. The second explanation originates from the terms "Heidenkorn" (German) and "Pohanka" (Polish), which translate to the cereal of fallows and the buckwheat produced by paynims. The third and final explanation is that the Spanish term "al-forfon vs. al-furfur" describes the dark red or purple color of the buckwheat plants [1]. In China, buckwheat is also called "black wheat," "flowering wheat," and "triangle wheat" [2].

Buckwheat is a cosmopolitan pseudocereal with limited production compared to cereals with higher yields. However, it is currently being introduced in many more countries, due to the high nutritional value of the buckwheat seed. The seeds are rich in essential amino acids, fatty acids, minerals, and vitamins $B_1$ and $B_2$. Buckwheat is also an important human dietary source. It is also used in honey production (possible yield is 120 to 300 kg of honey $ha^{-1}$); soon after the initial stages of flowering, two beehives $ha^{-1}$ can be placed. Every day, bees can collect up to 5 kg of honey from each beehive.

Buckwheat has a darker color and stronger taste than other kinds of flower honey, and it works as an antioxidant and plasma reducer in healthy human adults, potentially protecting the body from oxidative stress [3]. After bee pasture or unsuccessful fertilization, buckwheat can be plowed under and used as organic fertilizer: produce of above ground mass, which can also be used as quality animal feed, can amount to 7000 kg of dry matter ha$^{-1}$.

Buckwheat has become a popular food source all over the world, especially in Japan, the U.S., and some European countries, because of its use in traditional components of modern cuisine.

## 3.1.2   Botany

### 3.1.2.1   Taxonomy

This annual plant belongs to the *Polygonaceae* family and *Polygonum* genus. The *Polygonum* genus includes 15 species, three of which are known in common utilization: common buckwheat, known in in our case as simply buckwheat (*Fagopyrum esculentum* Moench, syn. *F. vulgare* Hill.); tartary buckwheat, sometimes called crazy buckwheat (*Fagopyrum tataricum* [L.] Gaertn., syn. *F. dentatum* Moench, *Polygonum tataricum* L.); and *Fagopyrum cymosum* Meins. Common buckwheat is divided into ssp. *vulgare* and ssp. *multiflorum* St., according to intensity of growth, number of leaves, and number of branches. *F. esculentum* ssp. *vulgare* includes var. *alata* Bat. and *aptera* Bat, according to characteristics of the seed var. *marginata* and *rhombea* [4].

### 3.1.2.2   Morphology

Cultivated and land race populations of buckwheat have various plant types. Certain standard differences exist between common and tartary buckwheat: tartary buckwheat has yellow-green flowers, wide and short leaves, and jagged margins of seed, while common buckwheat has petal flower leaves that are white or violet, without the characteristics of tartary buckwheat seeds. The depth of the main root varies between 0.6 and 1.2 m. Although roots are richly branched, the dry mass of root hair represents 3% of the full plant weight [4]. Root length increases constantly until peak flowering, reaching a length of 1.8 km m$^{-2}$ at harvest [5].

The stem often ranges from 0.3 to 1.0 m high, and can even reach 3.0 m, depending on genotype and growth conditions. The stem can either be branchless or branched with first order branches at the second, third, and fourth nodes.

In the upper inflorescence a growing point is hidden, which is active in suitable ecological conditions. High plant-population density can be used to prevent intensive side-branching. The leaves and stems of buckwheat are often red because of the presence of antocian.

A sphere appears over the upper-side branch, which begins to flower 4 to 6 weeks after sowing. Each flower cluster's blooming process lasts a relatively long time — about 20 days — and, consequently, ripening is not uniform (Figure 3.1). Forms displaying limited growth have been found

*Figure 3.1* Ripeness of buckwheat.

within some buckwheat genotypes; these genotypes are known as determinate buckwheat, an example of which are the Slovenian cultivars Siva in Darina. Determinate plants are more resistant to lodging and form more branches in suitable growth conditions.

The buckwheat's leaves grow from nodes. They are of cordate shape, 5 to 10 cm long, and alternately placed. The buckwheat flower consists of five petals of single flower integument, eight stamina, and one pistil. The flower formula is P5A5 + 3G(3), or P5G(3). Flowers are joined in clusters, and sometimes spikes form. Flowers in individual populations are divided into two types: those with long pistil and short stamina (pin type) and those with long stamina and small pistil (thrum type). Both types of buckwheat flower exist within one variety. Flowers are predominantly heterosterile; cultivars with a higher percentage of self-pollination are selected. In principle, the pollination of pin flowers is possible only with pollen from plants of the thrum type, and, conversely, pollen from the pin type may successfully fertilize only flowers of the thrum type [3].

The buckwheat seed is a single-seeded fruit: acorn (achene), three-edged seed, 6 to 9 mm long, with a rounded form. The acorn consists of a dicotyledonous embryo surrounded by an endosperm, enclosed by a testa and pericarp (hull). The embryo is found in the center of the endosperm, and cotyledons appear in the shape of the letters "s". One buckwheat plant can grow anywhere between 10 and 200 seeds. The seeds can be dark brown nearly black in color, or they may appear silver-grey to light grey. Hulls (per carp) comprise 18 to 40% of the acorn joint mass, which is determined by husking 20 samples of 20 acorns. The normal mass of 1000 acorns falls between 18 and 32 g, though it may reach as high as 38 g in metalloid form. Hectoliter mass varies from 54 kg hl$^{-1}$ to over 62 kg hl$^{-1}$.

The chemical components in an acorn are the following: 10 to 13% water, 61.4 to 66.9% extractive substances without nitrogen, 1.9 to 3.2% fat, 9.7 to 19.9% raw cellulose, 1.6 to 2.9% ashes, and 11.0 to 15.4% proteins [4]. The main phenolic components consist of epicatechin, rutin, hyperoside, and quartecin. A simple, reliable, and reproducible method based on capillary electrophoresis with electrochemical detection for their analyses has been found [6].

### 3.1.3   Production and yielding

Russia, China, and in 16 other countries currently produce 90% of the world's buckwheat across 3 million hectares of land. Other important buckwheat producing countries are Ukraine, Poland, Brazil, the U.S., France, Japan, Kazakhstan, and Canada. Smaller, traditional producers also exist in Hungary, Slovenia, Croatia, and the Baltic countries Lithuania, Estonia, and Latvia. Although data about the exact percentage of organically produced buckwheat is lacking, but it is clear that the organic production is increasing worldwide.

Grain yields are variable. If buckwheat is grown during a summer that is too hot or too cold, or in unsuitable soil or without pollination by insects, the yield may be as low as 500 kg ha$^{-1}$. The approximate yield for buckwheat is between 800 and 1000 kg grain ha$^{-1}$, but in favorable conditions, it can reach 2200 kg grain yield ha$^{-1}$ [4]. In some cases, producers have been satisfied using buckwheat as the main crop in areas with vegetation periods that exceed 150 days. Buckwheat grown as a full-season crop has a higher leaf area index, more clusters, better developed seeds, and 42% higher yield than the stubble-crop buckwheat [7]. Due in part to recent findings, the interest in buckwheat and its production is increasing, especially with regard to certified organic production.

### 3.1.4   Growth and ecology

Buckwheat seeds were often used as food in undeveloped and marginal regions. Due to higher yields of other field crops, however, the use of buckwheat did not spread to wider areas. The cultivation of buckwheat as a main crop was abandoned, according to globalization trends, because the production of main crops like corn and wheat that grew in monoculture proved more profitable [8].

Buckwheat demands specific climatic conditions. Production is possible far north (68° of northern latitude); in the Himalayas, buckwheat can even be produced at 4000 meters. Growth is limited by June isotherm 17°C in the north and 20°C in the south. It is very sensitive to cold: the lethal temperature for buckwheat ranges from –1.3 to –2.9°C, depending on the growth stage and conditions [9]. The plants are highly sensitive to frost from the early stage of primary leaves to the development of two secondary leaves. Differences among growth stages develop with plant hardening during growth and the critical time of frost action. Buckwheat is also not resistant to heat accompanied by drought. If temperatures exceed 30°C at the flowering stage, fewer flowers will form, and pollination will either be less satisfactory or simply not occur. In addition to oversensitivity to cold and heat, buckwheat is also vulnerable to profuse rain and wind; both may interfere with successful pollination. The decrease in grain yield may be caused by flooding at the flowering and the ripening stages for more than 10 and 3 days [10].

In order to achieve a growth capacity between 50 and 90%, soil moisture and temperature of 8–10°C is needed, with a minimal temperature of 4–5°C. The optimal germination temperature is 26°C; germination period at this temperature is 3 days. The optimal temperature during flowering is between 19 and 25°C. Assimilation stops with temperatures lower than 10°C; if temperatures exceed 24°C, pollination ceases [4].

Ripening of buckwheat (75% of acorns ripen) firmly correlates with accumulated growing degree days (AGDD, $r^2 = 0.93$). AGGD are also helpful for predicting optimal harvest time [11].

The transpiration coefficient of buckwheat is between 500 and 600. Although the transpiration coefficient of buckwheat is extremely high, the minimum amount of water necessary for buckwheat to survive, from germination to flowering, is 70 mm; at least an additional 20 mm are needed to reach the end of the growing period. A 3-day water stress during the first week of flowering reduces the number of seeds by 50% without reduction in seed size, dry weight, or number of formed flowers. The effect of water-deficit stress continues after irrigation and is expressed as a reduction in fertility and newly formed flowers [12]. Short days also hindered buckwheat growth. Plants under moderate irradiance (160 $\mu$mol m$^{-2}$ s$^{-1}$) exhibit higher growth, net assimilation rate, and osmotic adjustments than plants maintained under low irradiance (80 mumol m$^{-2}$ s$^{-1}$) [13]. Flower inductions also depend on day length and the number of inductive cycles. The critical day length varies with the genotype and growth parameters. Accordingly, it is necessary to choose the most appropriate cultivar for the main sowing or stubble-crop because such cultivar needs 9 hours of darkness for flower induction. According to some findings, the day before the start of anthesis significantly influenced the main stem elongation flowering process thereafter; this suggests that day length is a more critical factor for the differentiation than the growth of the flower bud [14].

Buckwheat growth peaks between 11:00 a.m. and 12:00 midday and is at its slowest between 6:00 p.m. and 9:00 p.m. It is significant for buckwheat to grow new shoots and leaves after flowering and even during ripening (though the determinant species is an exception). Buckwheat requires the most water in the first month, although the size of the leaf area multiplies later. Soil that is not plowed well does not enable uniform sowing depth, which results in uneven emergence, plant population, growth, and development [4].

The number of unfertilized flowers, formed seeds, open flowers, flower clusters, and ripening are conditioned by mutual influences, cultivar properties, sowing date, and plant population [15]. Branching largely depends on the cultivar and, according to numerous mutual influences on growth conditions, it is considered an important phenotypic property. Intensive branching can be regulated by sowing density with determinant cultivars, an important factor for reaching crop height [16]. Ripe acorns appear with buckwheat 5 to 10% of the time; an even more serious problem is caused by ripe acorns falling off. Buckwheat may take up to two months after the appearance of the first ripe acorns to harvest [17].

## 3.1.5  Organic cultivation practice

### 3.1.5.1  Crop rotation

Buckwheat production as a monoculture is not possible: continuous planting would result in diminishing yields. Buckwheat should not be planted again on the same field for at least 3 years, and 5 years is optimal. A negative consequence of incorrect buckwheat crop rotation is the increased occurrence of necrosis and the appearance of some disease signs, especially root diseases caused by *Gaeumannomyces* sp. When crop rotation is not correctly applied, more polyphagous insects will appear and the quantity of weeds increases. Buckwheat can be grown as a main crop or stubble crop after previous crops such as wheat, barley, early potato, and more, depending on the duration of growth season. Buckwheat presents high yields when grown following pea crops and fodder plants sown in autumn and harvested in the middle of the spring. The potato and buckwheat share the same nematode; nevertheless, the crop rotation includes potato and buckwheat and is still recommended. Buckwheat sown as a main crop in appropriate areas reaches high yield after other legumes or arable crops [4].

Buckwheat contains 10 types of allelochemicals (e.g., gallic acid, H-catechin, and so forth) that inhibit the growth of some weeds. The buckwheat pellets' selective inhibitory effects become greater in early application in transplanted rice [8]. Buckwheat may have allelopathic potential; when used as a ground cover crop or green manure, it may produce inhibitors that could suppress weeds [19] and cultivate plants.

### 3.1.5.2  Soil, plowing, and presowing preparations

Buckwheat can be produced in almost all types of soil, except in sand and wet or crusted soils. The optimal conditions are lighter, sandy-clay soils; products grown in extremely acid soil are of lower quality than those grown in neutral or lightly alkaline soil, but a high level of limestone in soil is also not suitable. Buckwheat should not be sown in soils rich in humus due to possible lodging.

Plowing stubble immediately is highly important to preserving moisture. After fertilizing, stubble should be plowed 12 to 15 cm deep. Fertilization can be undertaken immediately following plowing; in this case, quick presowing preparation is necessary to prevent any loss of moisture. Buckwheat must not be sown in dry soil [4].

### 3.1.5.3 Sowing and crop cultivation

There are 4500 known genotypes worldwide (cultivars and land race populations), mostly suitable for only a single climatic area due to their extreme adjustments to photoperiodic reactions (day length included). Therefore, their introductions into new climates represent a significant risk to successful production. Bavec et al. [7] suggest that the best-yielding buckwheat genotypes should be determined and introduced separately for stubble-cropor full-season production systems.

Several different sowing systems exist worldwide. Sowing is suitable with interrow spacing of 7 to 8 cm with row spacing of 5 to 6 cm. We usually sow buckwheat with interrow spacing of 10 to 12 cm, depending on how the seeder unit is set for sowing cereals. It is sown for grain with longer interrow spacing if the available seeded requires it — the interrow spacing should thus be doubled (to 24 cm), with very high sowing density, which increases the possibility of lodging. Buckwheat used as stubble crop is sown after barley, early potatoes, or early maturing cereals if the growing season is long enough to allow the maturity of the buckwheat. The sowing date of buckwheat as main crop is limited by spring freezing; sometimes, the plants will not flower, depending on the photoperiod of the genotype. Seeds should have at least 75% germination.

The depth of sowing in sandy soil is 3–4 cm and, in heavier soil, 1–2 cm. We use 60 to 90 kg of seed (acorns) ha$^{-1}$ when sowing with 250 to 300 viable seeds m$^{-2}$ [4]. Sowing between 750 and 1200 seeds m$^{-2}$ gives a higher yield than sowing 100 to 300 viable seeds m$^{-2}$ [15, 17]. University of Maribor results [29] show the increase of yield after an increase in sowing density to 1250 viable seeds m$^{-2}$ (sometimes as high as 1500 viable seeds m$^{-2}$), although with lower yield profitability, higher seed expenditure is not justified. Sowing too densely results in thin and less-branched plants with low productivity per plant. However, based on genotype and branching under specific climatic conditions, the preferred plant population ranges from 170 to 200 plants m$^{-2}$.

The provision of pollination is very important for the buckwheat crop: two beehives per ha$^{-1}$ are needed. As discovered by the University of Maribor, and also by Goodman et al. [21], the activities of honeybees and other insects doubled seed production.

Wide interrow sowing used in some areas around the world results in the possibility of cultivation between rows. Mechanical weed control is used only with wide sowing, because plants do not cover the soil quickly enough. As a rule, weed control is not needed, because the plant develops rapidly and suppresses summer weeds very effectively on its own [22]. Sowing healthy and clean seeds without weeds is critical.

Pests do not cause economically significant damage; therefore, buckwheat can easily be produced organically. Polyphagous insect problems can usually be traced back to unsustainable agricultural practices. Degeneration and damage may be caused by peronospora, grey mold, and the fusarium, but buckwheat is usually free from diseases. However, the economic assessment of damage is low and unknown.

### 3.1.5.4  Fertilization

In soil with low nutrient content, buckwheat grows better than other cereals. In humus soil with high nitrate content or potential mineralization, buckwheat often lodges. Buckwheat's demand for nutrients must be satisfied in order for it to be produced organically. According to our estimation, with production of 1500 kg of acorns per ha$^{-1}$ and 2500 kg of straw per ha$^{-1}$, the buckwheat crop uptakes about 45.5 kg of $P_2O_5$, 115 kg of $K_2O$, and 39 kg of CaO ha$^{-1}$. From this, grains take up 20 kg of $P_2O_5$, 15 kg of $K_2O$, 1.5 kg of CaO ha$^{-1}$, and about 27 kg of N ha$^{-1}$.

The required variation width of advised available nutrients differs in various production areas according to soil analysis, precipitation, and cultivars (from 44 to 88 kg N ha$^{-1}$ with lower production, to an unbelievable 234 kg N ha$^{-1}$ with production of 4000 kg acorns per ha$^{-1}$; from 26 to 132 kg ha$^{-1}$ $P_2O_5$ and from 33 to 165 kg ha$^{-1}$ $K_2O$). Sufficient boron content is necessary.

Nutrient uptake from organic fertilizers must be carefully considered and adjusted to meet the previously mentioned needs, with special attention paid to the precise uptake of nitrogen. If more than 20 g $NO_3$-N per kg$^{-1}$dry soil to the depth of 0.3 m exists during the sowing period, fertilization is not allowed. Mineralization will usually provide enough nitrogen, or the amount of nitrogen uptake does not exceed 40 to 60 kg N ha$^{-1}$. We do not use stable manure with stubble crop due to possible lodging problems; it should instead be used for the previous crop in the crop rotation. For basic or additional fertilization, compost is also suitable.

The efficient uptake of phosphorus from forms that present difficulty for other crops is a characteristic of buckwheat [4]. For example, it is highly efficient in taking up Ca-bound phosphorus compared to spring wheat, where the principal mechanism of phosphorus-uptake efficiency may be its ability to acidify the rhizosphere [23].

## 3.1.6  Harvesting, handling, and storage

Buckwheat is harvested when about 75% of the seeds have reached ripeness. Fully mature seeds are grey, brown, or black (depending on genotype characteristics) and fall off the plant during the vegetation period. The overall production period — from sowing to harvesting — should last 10 to 12 weeks under optimal climatic conditions. When buckwheat is sown as a full-season crop and the growing period exceeds 12 weeks, the decision about harvesting time depends on the number of fully mature seeds and the consequential grain yield. Humid and green masses cause difficulties at harvesting. The speed of the harvester and the rotation of the threshing cylinder have to be harmonized (600 rev. min$^{-1}$). High water content in plant stalks and seeds is typical at harvest (about 20-25%), whereas threshing during the dry summertime presents an exception. The grain should be dried to obtain 12% moisture (14% maximum) in the seeds for storage; otherwise, the seeds become stuffy and infected by mildews [4]. Edwardson [24] finds that it is safe to store buckwheat at 14 to 16% moisture.

Buckwheat may be stored under the same conditions as other grains. Due to the content of several natural antioxidants (tocopherols, flavanoids, and phenolic acids), buckwheat can be stored for long periods without any symptoms of chemical changes — only the occasional rancidity of grains or flour may occur.

## 3.1.7 Nutritional and health value

### 3.1.7.1 Nutritional value

Buckwheat seed contains between 11 and 15% high-quality and easily digestible proteins, which represent more than 90% of the protein value in skimmed milk and more than 80% of the protein value in dry eggs. Most represented proteins are globulins (approximately 40%), whereas prolamin content is low. Buckwheat grain contains 1.5 to 3.7% of lipids. Oils contain 16 to 20% of saturated fatty acids, 30 to 40% of oleic acids, and 31 to 41% of linoleic acid [25].

The nutritional value of buckwheat seeds and the products of milled seeds depend on the relative abundance of the various seed tissues in each (Table 3.1).

*Table 3.1* Protein, Carbohydrate, Oil, Fiber, and Ash Content of Buckwheat Milling Fractions Compared with Other Products (% in dry matter)

| Product | Protein | Carbohydrates | Fat | Fiber | Ash |
|---|---|---|---|---|---|
| Seed | 12.3[a] | 73.3[a] | 2.3[a] | 10.9[a] | 2.1[a] |
| Whole groats | 12[c] | 57[c] | 4[c] | 7[c] | 2[c] |
| Bran | 36[c] | 24[c] | 11[c] | 15[c] | |
| Groats | 16.8[a] | 67.8[a] | 3.2[a] | 0.6[a] | 2.2[a] |
| Dark flour | 14.1[a] | 68.6[a] | 3.5[a] | 8.3[a] | 1.8[a] |
| Average flour | 11.7[a] | 72.0[a] | 2.5[a] | 1.6[a] | 1.8[a] |
| | 11.6[b] | | 2.3[b] | | 1.8[a] |
| White flour | 6.4[a] | 79.5[a] | 1.2[a] | 0.5[a] | 0.9[a] |
| Wheat flour | 11.8[a] | 74.7[a] | 1.1[a] | 0.3[a] | 0.4[a] |

[a] [26].

[b] Adapted data from [27].

[c] [28].

*Source:* Robinson, R.G., The Buckwheat Crop in Minnesota, *Agr. Exp. Sta. Bul.*, Univ. Minnesota, St. Paul., 539 pp. With permission from University of Minnesota Extension Service, published in 1980.

*Source:* From Belton, P. and Taylor, J., *Pseudocereals and Less Common Cereals: Grain Properties and Potential*, 2002. Copyright 2005, with permission of Springer Science and Business Media.

*Source:* From Steadman, K.J., Minerals, Phytic Acid, Tannin, and Rutin in Buckwheat Seed Milling Fractions. Copyright Society of Chemical Industry. Reproduced with permission. Permission is granted by John Wiley & Sons Ltd. on behalf of the SCI.

*Table 3.2* Essential Amino Acid Composition of
Buckwheat, Wheat, and Corn

| Amino Acid | Buckwheat | Corn | Wheat |
|---|---|---|---|
| Isoleucine | 3.57–3.69[a], 0.46[b] | 0.33[b] | 4.42[a], 0.46[b] |
| Leucine | 5.95–6.29[a], 0.84[b] | 1.20[b] | 8.24[a], 0.91[b] |
| Methionine | 0.97–1.14[a], 0.19[b] | 0.18[b] | 1.02[a], 0.18[b] |
| Phenylalanine | 4.11–4.54[a], 0.56[b] | 0.47[b] | 5.17[a], 0.64[b] |
| Threonine | 3.32–3.61[a], 0.49[b] | 0.33[b] | 3.12[a], 0.40[b] |
| Valine | 5.31–5.43[a], 0.60[b] | 0.44[b] | 5.12[a], 0.56[b] |
| Lysine | 5.38–5.76[a] | – | 3.26[a] |

[a] [27] (in mg g$^{-1}$ dry matter).

[b] [29] (%).

*Source*: From Belton, P. and Taylor, J., *Pseudocereals and Less Common Cereals: Grain Properties and Potential*, 2002. Copyright 2005, with permission of Springer Science and Business Media.

The percentage of essential amino acids in the entire amount of amino acids ranges from 8.6 to 9.3. The essential amino acid composition of buckwheat (Table 3.2) is more balanced and of better nutritional value than that of other cereals [25, 29]. Buckwheat is especially rich in lysine [27] and some minerals [30]. Leucine is the first limited amino acid [31].

Buckwheat is generally rich in vitamins and essential minerals. It is an important source of thiamine (vitamin $B_1$), riboflavin (vitamin $B_2$), and other vitamins from the B group, of which buckwheat acorns contain 150% more than wheat grain. Buckwheat contains natural antioxidant tocopherols (vitamins from the E group) and is also rich in vitamin P, which is found primarily in flavonoid rutin and is used to stop capillary bleeding [4].

Buckwheat mineral content is also favorable; it contains Fe, Zn, Mg, P, and more (Table 3.3) and 0.039 to 0.053 mg Se 100 g$^{-1}$ of seeds.

The buckwheat seed is rich in K and Zn in the albumin; Ca, Mg, and Mn in the globulin; and Na in the prolamin and glutenin [31].

### 3.1.7.2    Health value

Buckwheat is used as a dietary food for children and the ill; its nutritional and medical properties make it a useful and interesting seed to many people.

Buckwheat is also a highly effective nutritional food. Besides high-quality proteins, buckwheat seeds contain several components with healing benefits: flavonoids and flavones, phytosterols, fagopyrins, and thiamin-binding proteins [32]. Buckwheat product consumption can improve diabetes, obesity, hypertension, hypercholesterolemia, and constipation [33]. It can also prevent gangrene, frost bites, and strengthen the human organism after radiation. Buckwheat has been shown to reduce the caloric value of meals and preserve blood sugar at the optimal level [4]. Food from buckwheat flour is not toxic for people with celiac disease, because it does not contain protein gluten [34]; thus, it is often appreciated for its culinary properties.

*Table 3.3* Comparison of Mineral Content Among Buckwheat, Corn, and Wheat

| Product | Amount (mg 100 g⁻¹) | | | | | | | | |
|---|---|---|---|---|---|---|---|---|---|
| | Ca | Fe | Mg | P | K | Na | Zn | Cu | Mn |
| Buckwheat[a] | 11.6–11.0 | 17.5–4.0 | 173–390 | 426–330 | – | – | – | – | 14.3–3.37 |
| Buckwheat[b] | – | – | – | – | – | – | 22.5–37.5 | 5.3–7.2 | 17.5–22.4 |
| Buckwheat flour[a] | 41 | 4 | 251 | 337 | 577 | 0 | 3.12 | 0.09 | 0.099 |
| Corn flour[c] | 6 | 3,45 | 127 | 241 | 287 | 35 | 1.82 | 0.193 | 0.498 |
| Wheat flour[c] | 15 | 1,17 | 22 | 108 | 107 | 2 | 0.70 | 0.144 | 0.682 |

[a] [7].

[b] [30].

[c] [24].

*Source*: From Belton, P. and Taylor, J., *Pseudocereals and Less Common Cereals: Grain Properties and Potential*, 2002. Copyright 2005, with permission of Springer Science and Business Media.

*Source*: Reprinted from Bavec, F., Pusnik, S., and Rajcan, I., *Rostlinna Vyroba*, 48 (8), 351–355. Copyright 2005, with permission from Rostlinna Vyroba.

*Source*: Reprinted from *Progress in New Crops*, Edwardson, S. and Janick, J., Eds., 1996, 195. Copyright 2005, with permission from ASHS Press.

As mentioned earlier, buckwheat is rich in vitamin P, a component of flavanoid rutin. Rutin has an anti-inflammatory, hypertensive effect; it reduces fragility of blood vessels associated with some coronary diseases [35], binds estrogenic receptors, and has antitumoral properties [36]. Flavanoids such as rutin and isovitexin have been found in groats, while hulls also contain isorientin, orientin, quarcetin, and vitexin [37]. Since flavanoids have an antioxidant effect, foods made with buckwheat can be used to prevent diseases. A buckwheat diet reduces blood pressure [38], reduces concentration of cholesterol in serum, helps with activity of the hamster's liver [39] and gallbladder, and suppresses the formation of gallstones [40]. Similar effects upon cholesterol, metabolism, and hypocholesterol were described by humans, most likely due to the soluble fiber content in buckwheat foods [41].

Prestamo et al. [42] conclude that buckwheat is a prebiotic and healthy food, because higher levels of lactic acid bacterias like *Lactobacillus plant arum* and *Bifidobacterium lactic Bifidobacterium* ssp. were found in buckwheat diets than conventional diets. By definition, prebiotic refers to a nondigestible food ingredient that beneficially affects the host and selectively stimulates the growth and activity of one (or a limited number of) bacteria in the colon. Asian countries typically have lower numbers of cardiovascular accidents and cancers than the occidental countries, and research on the subject [42] mainly attributes these numbers to the buckwheat diet, like the soba noodles commonly eaten in Japan. Deschner [43] and Liu [44] also report that buckwheat reduces cellular proliferation and therefore protects the colon against carcinogenesis.

Buckwheat has been described in recent scientific articles as a plant that causes various allergies, as a consequence of hypersensitivity to foreign proteins (anaphylaxis) [45]. Asthma attacks, gastrological problems, cold, eye problems, epileptic fits, and similar symptoms are mentioned most often. Food that contains fagopyrin may indeed cause skin inflammation (urticaria) in some people. But allergies may also be transferred through the air by thermostable proteins with high molecular mass. These proteins can cause so-called "baker's asthma," because flours are powders of varying degrees of fineness [46].

### 3.1.7.3  Processing

Products made from buckwheat are flour, hulled grains (groats often called kasha in English and sometimes known as buckwheat rice in Japan), processed dry, frozen, and instant buckwheat (soba) noodles, and hulls for pillows. Flour can be used as an additive when making chocolate, cream, cakes, canned meat and vegetable products, and breakfast cereals (Figure 3.2). In some countries, buckwheat is also used to prepare alcohol drinks. A developing technology applies the process of extraction to the production of snacks containing buckwheat flour. Buckwheat could be used to an even greater extent by processing cereals from hulled grains.

Buckwheat's technological value (thousand achene's weight, volume weight, proportion of fractions on sieves 4.5 and 4 mm, proportion of husks, and percentage of groats) can be influenced by weather (particularly rainfall) during flowering and achene formation periods [47], and also by genotype.

### 3.1.6.7  Milling

Flour processing is done by hand or by water-driven stone mills and grinding stones in buckwheat-growing countries. The old hand-driven or water-driven mills do not develop excessively high temperatures during the milling process, which results in the typically pleasant buckwheat taste; quick milling to flour, in contrast, causes long exposure to high temperatures, and the good taste is lost. Two main flour types are produced: the first from whole grains with separations of different flour fractions; and the second from hulled grains, where several types of broken endosperm are obtained as inner-layer endosperm, further classified in types. The characteristics of flours (flavor, masticating, overcooking, and so forth) will vary, depending on the flour type [2]. Fine (light) flour mainly contains endosperm. It is possible to get bran milling fractions from the seed coat and some embryo tissues [28]. It is also possible to separate 60 to 70 kg of flour of standard darkness quality from 100 kg of grains during milling; what remains are scales and bran. The production of fine (light) flour does not exceed 52%.

Traditional processing of buckwheat groats (kasha) (Figure 3.3) is based on dehusking grains after thermal treatment (mostly cooking and backing),

*Figure 3.2* Buckwheat flakes.

*Figure 3.3* Buckwheat groats (kasha).

similar to the autoclaving process. At high temperatures, drying grains start to open and need to be extruded mechanically, with the use of rough metal surfaces or rotating disks. New groats processing is based on (i) thermal treatments of grains or autoclave hulling in combination with high pressure or (ii) natural hulling of raw and dry grains (named "cold" processing). "Cold" processing is based on special, mainly secret technical solutions.

New food products can be developed from buckwheat by using modern hydrothermal technologies such as flaking, extrusion, and puffing. In bakery, pasta, and confectionary products, buckwheat flour does not contain any gluten; therefore, the elasticity and plasticity characteristics of its dough are not comparable to dough made from wheat flour [27]. As mentioned previously, a maximum of 30% buckwheat flour is recommended in a mixture for bread making. In producing confectionary products such as cakes, biscuits, and crackers, the lack of gluten and the rheological parameters of buckwheat do not cause any significant problems. Processing conditions — including tempering moisture, heating temperature, and heating time — significantly influence the physical and chemical qualities of buckwheat grit cakes, such as the volume, hardness, integrity, color, internal structure, and rutin content [48]. Buckwheat grain used to occasionally be distilled into alcohol, but today it is used in breweries for the production of high-quality beer.

## 3.2   Quinoa

### 3.2.1   Introduction

Quinoa, also known as quinua or white quinoa (*Chenopodium quinoa* Willd.),is becoming a more and more interesting organic food crop [49] due of its high nutritional value; it provides an exceptional combination of vitamins, minerals, high protein quality, and essential amino-acid composition. Quinoa is a native plant to the Andes Mountains. The grains and young leaves of quinoa have been consumed for many years by the people of the mountain regions of Bolivia, Chile, Ecuador, and Peru [49, 50]. In the Inca language, quinoa means "mother grain"; the quinoa plant has been an important staple food for centuries.

Quinoa cultivation began to decline in Andean countries with the development of intensive agriculture. The introduction of cereals such as barley and wheat brought cheaper products, due in part to mechanized harvesting and threshing processes. The cost to the farmers who were growing quinoa often did not justify their labor; thus, barley and wheat eventually replaced quinoa. At present, quinoa continues to be grown in Colombia, Ecuador, Peru, Bolivia, Chile, and Argentina.

Quinoa has been selected by FAO (1998) as one of the crops destined to solve the food security problems of the Andes during the twenty-first century, due to its nutritional qualities, cultural acceptability, increasing production, and marketability. The main producers and exporters of quinoa grain to Japan, Canada, and the EU are Bolivia and Peru, where the production of quinoa is still increasing. Bolivia produces an average of 47534 ha and 640 kg grain yield ha$^{-1}$, and Peru averaged 28355 ha with 920 kg grain yield ha$^{-1}$ in 1999. In recent years, the export price for grain has varied between 0.56 and 1.77 USD kg$^{-1}$, and the price for organically produced yield exported to Europe was 3.00 to 3.50 USD libra$^{-1}$ (lb = 0.45 kg). Quinoa is exported to the U.S. from the provinces of Buenos Aires and Córdoba in Argentina; around 500 hectares of quinoa are also grown within the U.S. (Colorado) as a commercial crop [51*].

* With permission from the Food and Agriculture Organization of the United Nations (FAO), authoroziation number A138/2005; Mujica, A. et al., Quinoa (*Chenopodium quinoa* Willd.): ancestral cultivo Andino, alimento del presente y futuro, project FAO presentation, Santiago, Chile, 2001, 303 pp.

Despite the extreme reduction in quinoa production, the interest in this food crop increased after 1975 and was experimentally produced in 22 countries outside South America until 1989, especially as fodder plants for animals in Europe. Some European countries, such as Great Britain, Spain, Denmark, Germany, Finland, and Slovenia, are studying its adaptation, breeding, and possibilities for commercial production [50].

## 3.2.2 Botany

### 3.2.2.1 Taxonomy

Quinoa (*Chenopodium quinoa*, tetraploid, 2n = 36) is an annual herbaceous plant and belongs to the family *Chenopodiaceae* (*C.* species may be diploid, tetraploid, hexaploid, or octoploid); beet (*Beta vulgaris* L.) and spinach (*Spinacea oleracea* L.) belong to the same family. The genus *Chenopodium* consists of 150 species, classified into 16 sections. Few plants from section *Chenopodium* are used for human consumption. Besides quinoa, *Chenopodium pallidicaule* Heller is produced in Peru; *Chenopodium berlandierri* Moq. ssp. *nuttalliae* Safford is produced in Bolivia; and *Chenopodium album* L. is cultivated in Himalaya as food crops. *C. album* (hexaploid) and *C. berladieri* are more known as weeds. *C. berlandieri* can cause breeding problems, because it is tetraploid and crosses readily with quinoa. Quinoa is the most important species from genus *Chenopodium*. In the first stage, cultivated genotypes originated from 6 ecotypes and 4 plant populations from different regions [50].

### 3.2.2.2 Morphology

Cultivated genotypes of quinoa exhibit great genetic diversity. Many differences exist among genotypes in plant height (from 0.2 to 3.0 m), the coloring of the plant, florescence and seeds, degree of stem branching, leaf morphology, types of inflorescence, seed weight, protein content, saponin content, calcium oxalate crystals in the leaves, and so forth. Most vegetative, floral, and yield characteristics are controlled by genetic factors, but they are also affected by environment and production.

Quinoa has a semivigorous and deep-rooting taproot with branched secondary and tertiary roots. The root depth and plant height represent equal portions. After emergence, the root elongation of young plants occurs quickly in the presence of adequate temperature and soil moisture. Ten weeks after sowing, the taproot depth remains shallow, perhaps reaching a depth of 30 cm, but this characteristic varies among genotypes. The root branches are formed just bellow the rootneck. Deep roots and branching may be one of the reasons for the high degree of drought resistance among quinoa plants.

Quinoa plants have upright stems, which form branched or unbranched plants, depending on genotype. In branched genotypes, the number of branches depends on climatic conditions and density of plant population. A high degree of branching is not undesirable for grain production, but an adequate grain yield is reachable under optimal relationships among plant population and plant branching for each growing condition. The lengths of branches, which originate from the axils of each leaf on the stem, vary few centimeters in length between the axils of the leaf and the top of the main stem or inflorescence. The outside of the stem and branches can be green, red, or yellow, with different-colored stripes, or green with colored axils (red or purple). In the early growth stages, the leaves are covered with granular bladdery excrescence, which passes on to white and purple color. A mature plant is pale yellow or red to purple in color. Leaves have different leaf laminae; lower laminae are mostly rhomboidal, while the upper leaves are triangular or lanceolate. Most often, the lamina is plane, but in some genotypes it can be undulating. The number of leaf serratations may vary from 0 to 20 teeth; in serrated genotypes, the upper leaves have fewer teeth (serratations). Leaf area varies depending on the position of the leaves on the plant, the genotype, and the plant population. Leaf area of middle leaves range from 19.0 to more than 50.0 cm$^2$, and between 3.0 and 10 cm$^2$ for upper leaves.

Quinoa has a recemose inflorescence (panicle) containing many different colors. Its branching depends on genotype (Figure 3.4), but in general, two types of inflorescences exist: lower inflorescence, or amaranthiform, where flowers (glumeruli) originate from secondary axes; and upper inflorescence, or glomerulate type, where groups of flowers originate from tertiary axes. The flowers are incomplete without petals. Quinoa is highly self-fertile, because most of its flowers are hermaphroditic, but they may also be female

*Figure 3.4* Branches of quinoa.

(pistillate) or male sterile (usually with hybrids). The size of hermaphroditic flowers varies from 2 to 5 mm, and females grow up to 3 mm. Hermaphroditic flowers contain a five-numbered perigonium, a pistil, and an ellipsoid ovary, and the stamens have short filaments bearing basifixed anthers. Additionally, the style has two or three feathery stigmas. The female flowers have only perigonium and pistil [52].

The fruit is an achene, protected by the perigonium. Different colors in the inflorescences are caused by the coloration of the perigonium. The seeds are yellow, red, purple, brown, black, or white; colors are determined by the color of the pericarp, or by the episperm when the pericarp is translucent. The quinoa seed has an unusual structure; a medial longitudinal section showing the pericarp, hypocotyl-radicle, and cotyledons; endosperm in the micropylar part only; and radicle, funicle, shot apex, and perisperm [53]. Seeds vary in weight from 2 to 6 mg and in size from 1 to 2.6 mm, and they may be sharp, conical, cylindrical, or ellipsoidal. The seed mainly possesses sharp borders, but it can also have a rounded border.

### 3.2.3   Growth and Ecology

According to summaries of different descriptions [51, 54, 55] and personal notices, the phenological stages of quinoa are:

1. Vegetative (V): Seed germination and emergence, true leaves formation (two, four, or six true leaves), and branching (first order branches at the second, third, and fourth nodes); and
2. Reproductive (R): Bud formation (covered, visible, and distinct in cm, shape), florescence or anthesis (onset, 50% and 100% flowering), and seed formation and maturity (woody seed, soft seed, or full mature seed).

The vegetation period from sowing to harvesting may last anywhere from 120 to 240 days [55], depending on genotype and growth stages, with the stage from two true leaves to bud formation lasting between 41 and 89 days; from bud formation to anthesis lasting between 7 and 53 days; and from anthesis to maturity lasting between 65 and 137 days [56]. The durations between the true leaves stage and bud formation and between seed formation and full maturity require the longest times. There are no significant differences among genotypes. Quinoa is a long-day plant, with significant influences on duration of vegetation periods. An optimal supply of nutrients may influence the transformation into vegetative stage even more [57].

Quinoa adapts easily to different topography and altitudes. It can be grown in adverse environmental conditions such as cold, soil salinity, drought, and humidity [50]. In dry years, water shortage may cause considerable yield reductions, but a minor draught stress does not result in a large yield decrease [58]. Dry periods in the true leaves stage can result in significantly smaller surfaces of side leaves, lower green and dry matter weights

*Figure 3.5* Research of different irrigation treatments affected growth in quinoa plants (from the University of Maribor).

of all leaves, lower inflorescence weights, lower root weights, and less branching in comparison to optimal conditions. Plants that follow later can partially compensate for a lack of humidity, which is a benefit for producers in dry areas (Figure 3.5) [59]. A noticeable production increase appeared in plants subjected to drought during the branching stage than during other growth stages [58]. Periods from full-field water capacity to complete dehydration of soil, when all water available to plants was used, varied according to the development stages of quinoa. The period was the longest at branching stage — where it lasted 16 days — and the shortest at flowering stage — where it lasted 6-days; a 10-day period was noted at the grain-filling stage [60].

Research shows that misestimations and poor knowledge regarding quinoa water demands caused damage, due to the overirrigation of plants. A lower irrigation degree of quinoa plants is recommended for the Altiplana region, whereas for cultures sensitive to drought and with higher water demands, increased irrigation is recommended [58]. Plants are most sensitive to cold in the early growth stages and can survive from –3 to –15°C, depending on the genotype's ontogenic resistance. Plants with red leaves are less sensitive to cold [50]. Acceptable production is also obtained in acid soils with a pH of 4.5 and alkine soils with a pH of up to 9.5, common for some Peruvian and Bolivian ecotypes, respectively. Production appears both in sandy and clay soils, but it prefers sandy-loam to loamy-sand and semi-deep soils with good drainage and nutrient supply. However, marginal agricultural soils, which have low natural fertility, poor drainage, and high acidity, are frequently used to grow quinoa [61].

## 3.2.4 Organic cultural practice

### 3.2.4.1 Crop rotation, pests, and diseases

With balanced biodiversity and suitable crop rotation, avoiding plants from the family *Chenopodiaceae*, quinoa should not require special pest and disease control. However, viruses found on spinach or beets have been observed in quinoa fields. Diseases caused by *Peronospora* sp., *Sclerotinium* sp., *Phoma* sp., *Botrytis* sp, and *Pseudomobas* sp. may have an influence on yielding [61]. Several species of cosmopolitan polyphagos insects (e.g., *Agrotis ipsilon* Hufnagel, Lepidoptera: nuctuidae) may cause economic losses in quinoa production. Some of widespread pests endemic to the Andean region may cause a 15 to 50% loss of yield, especially two species of moths (*Eursacca quinoae* Povolyny, *E. melanocampta* Meyrick) [62], which is a cause for developing new strategy of research [63]. Emerging plants may also be damaged by flea beatles and caterpillars. Quinoa is generally cultivated in rotation with potatoes and cereals. The quinoa seeds do not exhibit dormancy and they germinate when conditions are suitable; the plant itself though, in wild form, may remain in soil for 2 to 3 years without germinating [55].

### 3.2.4.2 Sowing and intercultural operations

Grain yield appears to strongly depend on sowing density and date [49, 56, 64]; however, the recommended plant population varies a great deal worldwide. The best time for sowing and the rational sowing rate depend primarily on the climatic conditions prevailing in the given area. Applied sowing rate and the interrow and row distances also depend on the available mechanization for sowing, weeding, and cultivation. Quinoa can be sown early in spring when the soil temperatures range from 5 to 7°C; but plants grow slowly during the first 2 weeks following emergence. Early planting may contribute to effective control of pigweed and other summer weeds, since quinoa will have a good head start on growth before the weeds emerge.

Many different methods of sowing are used in quinoa-producing countries. The current recommendation calls for wide-row sowing with interrow spacing of 0.2 to 0.4 m or manual sowing in heaps of five seeds 0.4 m in diameter, and a distance between heaps of 0.7 to 1.2 m. According to some studies, the most suitable interrow spacing is 0.2 m with 20 kg of seed ha$^{-1}$. In this case, plants cover the soil surface within 45 days of sowing; this makes it possible to achieve optimal plant distribution and prevent high weed competitiveness [56]. Such sowing methods are usually less suitable due to a lack of weeding mechanization, especially in early growth stages. Interrow spacing should be adjusted to the available mechanization and can range from 0.4 to 0.7 m. An amount of seed higher than 20 kg of seed ha$^{-1}$ resulted in shorter plants; lower seed amounts resulted in the high branching of plants and earlier maturity. In South America, the suggested amount according to production conditions and available mechanization is 4 to 6 kg of seed ha$^{-1}$; elsewhere, 8 to 12 kg of seed ha$^{-1}$ or more are recommended. Mujica [55] suggests 15 to 20 kg unselected seed ha$^{-1}$. In humid conditions, plants may be weak with low seed production. The number of plants in experiments varies from 30 plants m$^{-2}$ [65] to an optimal plant population of 140 plants m$^{-2}$ in a greenhouse with controlled climatic and hydroponical conditions [66], to 200 plants m$^{-2}$ in the field trials [52]. For cultural practice, 100 to 150 plants m$^{-2}$ are suggested in the U.S. and about 800 seeds m$^{-2}$ to 2000 seeds m$^{-2}$ are suggested in South America [66].

Cultivation work, especially in furrows spaced 0.4 to 0.8 m apart, is limited to one or two hoeings during the first phonological phase. Organic fertilizers that are left over from the preceding crop or given stable manure in autumn provide adequate yielding in the organic production system. Due to lack of data about organic fertilization in quinoa, compensation is suggested by organic fertilizers at an application rate of about 120 kg of N ha$^{-1}$ [65, 67] for the expected enhancement of grain yield.

## 3.2.5   Harvesting

Quinoa is physiologically mature when grains are dry and firm and we are unable to thresh them in hand. At the maturity stage, plants become lighter and more yellow, and the leaves fall off. Traditionally, plants were pulled or cut, and whole plants were dried. Now, the crop can be harvested using a combine; however, this should only be conducted with dry plants. There is no major loss of seed noted at harvest. A lot of damage can be caused by birds; in some test fields, they picked approximately 40% of produce. Usual harvest is 400 to 1200 kg grain yield ha$^{-1}$. Under experimental conditions, harvest reached as high as 435 to 6591 kg of grain yield ha$^{-1}$ [50].

## 3.2.6   Chemical composition, nutritional and health value

The proximate composition of quinoa ranges from 10 to 18% for crude protein, from 54.1 to 64.2% for carbohydrates, from 4.5 to 8.75% for crude fat, from 2.1 to 4.9% for crude fiber, and from 2.4 to 3.65 for ash [68]. Compared data (Table 3.4) shows wide ranges, which are probably a reflection of differences among genotypes and growing conditions and their effects on protein-fat ratios in perisperm. However, the protein content is slightly higher [69], to 5% higher [70], than that of most other cereal grains. Some cultivars from Ecuador are very protein-rich. Additionally, the fat content is at least twice as high as in most cereals. Seed content comprises approximately 60% of starch from the total carbohydrates, and the remaining are mostly free sugars. Carbohydrates in quinoa seeds display some important physical and chemical properties, especially smaller starch grains within the size of 1 to 2 μm in comparison to corn (1 to 23 μm) and wheat (2 to 40 μm). Due to amylases and a high capability of water uptake, starch is extremely viscous and good for making starch paste [71*].

---

* European Patent Application, No. 891216554.1, 1989, With permission from European Copyright Office.

*Table 3.4* Proximate Composition of Quinoa (%) According to
Different Authors [72–74]

| Component | Fleming and Galwey [72] after Romer | Koziol [73] | Ruales [74] |
|---|---|---|---|
| | (Variation) | | |
| Moisture | 12.9 (5.4–20.4) | 9.6 | Dry |
| Protein | 14.3 (9.6–22.1) | 15.7 | 14.1 |
| Starch | 61.4 (46.0–77.4) | 61.7 | 51.6 |
| Fat | 4.6 (1.8–8.2) | 7.2 | 9.7 |
| Fiber | 3.0 (1.1–5.8) | 2.9 | 13.4 |
| Ash | 3.5 (2.4–9.7) | 3.3 | 3.4 |

*Source*: Reprinted from *Underutilized Crops: Cereals and Pseudocereals*, Fleming, J.E., Galwey, N.W., and Williams, J.T., Eds., Chapman & Hall, London, 1995, 3–85. Copyright (2006), with permission from Kluwer Academic Publishers.

*Source*: Reprinted from Koziol, M.J., *J. of Food Composition and Anal.*, 5, 35–68. Copyright 2005, with permission from Elsevier.

*Source*: Reprinted from Ruales, J., Development of an Infant Food from Quinoa (*Chenopodium quinoa*, Willd.): Technological Aspects and Nutritional Consequences, Ph.D. Thesis, University of Lund, Lund, Sweden. Copyright 2005, with permission from LUND University.

Proteins in quinoa seeds are mainly of the globulin and albumin types, without gluten fraction. Amino acid composition (Table 3.5), except for leucine, meets the ideal composition reference pattern for children, according to the FAO/WHO/UNU standards; additionally, 100 g of quinoa seeds meet 200% of adults' daily need for histidine, 337% of the need for isoleucine, 347% of the need for lysine, 312% of the need for methionine and cystine, 363% of the need for phenylalanine, 411% of the need for threonine, 180% of the need for tryptophane, and 346% of the need for valine.

Quinoa is rich in lysine, the first limiting essential amino acid in cereals. Seeds also contain high amounts of histidine, and the value of isoleucine and methionine + cystine is higher in comparison with other cereals [73].

The mineral content of quinoa grains varies, depending on cultivars and growing circumstances, but they are overall rich in phosphorus, magnesium, and calcium (Table 3.6), similar to cereals. The mineral content in the hull grains is a few times higher than in the dehulled grains and differs among milling fractions.

According to the recommendations of dietary allowance, quinoa can be classified as a source of tocopherols (vitamin E), riboflavin ($B_2$), thiamin (vitamin $B_1$), and folic acid. Unlike cereals, quinoa also contains vitamin C [69]. The content of alpha-tocopherol varies from 2.0 to 5.4 mg 100 $g^{-1}$ grain, thiamine from 0.3 to 0.4 mg 100 $g^{-1}$ grain, riboflavin from 0.2 to 0.4 mg 100 $g^{-1}$ grain, about 78 μg 100 $g^{-1}$ grain, and from 3.0 to 16.4 mg 100 $g^{-1}$ grain of vitamin C [78, 79]. The Ca:P ratio in quinoa is more suitable (1:1.25) than with cereals (1:10.7), considering thatthe optimal ration is 1:1.

Quinoa contains a less appropriate component in the pericarp of its seed, from a nutritional point of view, known as saponin. Saponin is a detergent

*Table 3.5* Essential Amino Acid Composition of Quinoa (g 100 g$^{-1}$ protein) According to Different Authors

| Amino Acid | Modified by Telleria et al. [75] Variation Among Four Cultivars | Koziol [73] | Becker and Hanners [76] |
|---|---|---|---|
| Histidine | 2.1–2.8 | 3.2 | 2.6 |
| Isoleucine | 2.4–3.8 | 4.4 | 3.7 |
| Leucine | 5.9–7.4 | 6.6 | 5.9 |
| Lysine | 5.1–6.7 | 6.1 | 5.6 |
| Methionine+cystine | 11.2–21.0 | 4.8 | 3.8 |
| Phenylalanine+ Tyrosine | 4.8–6.3 | 7.3 | 6.6 |
| Threonine | 2.9–4.1 | 3.8 | 3.5 |
| Tryptophan | 0.8–1.0 | 1.1 | – |
| Valine | 3.5–4.8 | 4.5 | 4.9 |

*Source:* Reprinted from *Archivos Latinoamericanos de Nutrition,* 28, Telleria, M., Sgrabieri, V.C., and Amaya, J., Evaluación quimica y biológica de la quinoa (Chenopodium quinoa Willd). Influencia de la extracción de las saponinas por tratamiento térmico, 253, Copyright (2006), with permission from ALAN.

*Source*: Reprinted from Koziol, M.J., *J. of Food Composition and Anal.,* 5, 35–68. Copyright 2005, with permission from Elsevier.

*Source:* Reprinted from Becker, R. and Hanners, G.D., *Food Sci. and Technol.,* 23, 441–444. Copyright 2005, with permission from Elsevier.

*Table 3.6* Mineral Content in Quinoa Grain and Milling Fractions

| Mineral | Whole Grain[a] | Whole Grain[b] | Whole Grain[c] | Flour[c] | Bran[c] |
|---|---|---|---|---|---|
| Phosphorus | 360 | 535 | 470 | 320 | 670 |
| Calcium | 110 | 87 | 190 | 150 | 240 |
| Potassium | 900 | 120 | 870 | 470 | 160 |
| Magnesium | 500 | 262 | 260 | 160 | 460 |
| Natrium | – | – | 1.1 | 6.7 | 0.6 |
| Iron | 9 | 8 | 2.0 | 1.8 | 2.5 |
| Copper | 1 | 1 | 0.7 | 0.6 | 0.8 |
| Manganese | 4 | 3 | 1.3 | 1.2 | 1.6 |
| Zinc | 0.8 | 3.6 | 0.5 | 0.5 | 0.8 |

[a] Chauhan et al. [77].

[b] Ruales and Nair [78].

[c] adapted by Taylor et al. [69] .

*Source*: Source: Reprinted from *Cereal Chem.,* 69, Chauhan, G.S., Eskin, N.A.M., and Thachuk, R., Nutrients and antinutrients in quinoa seed, 85, 1992, Copyright 2006, with permission from American Association of Cereal chemists.

*Source*: From Belton, P. and Taylor, J., *Pseudocereals and Less Common Cereals: Grain Properties and Potential,* 2002. Copyright 2005, with permission of Springer Science and Business Media.

*Source*: Reprinted from Ruales, J. and Nair, B.M., *Publ. Food Chem.,* 48, 131. Copyright 2005, with permission with Elsevier.

soluble in water and is found within plants in approximately 15 molecular forms. Saponin removed from grains represents natural substances used in organic soaps, detergents, shampoos, cosmetics, and so forth. These types of products may avoid water pollution after grain washing in the first step of food processing because some saponins are known to damage the intestinal membrane and reduce food consumption. However, saponin reduces the amount of cholesterol in the blood of people and birds. Some isolated saponins, found in many plants, may be used as natural insecticides that are harmless to mammals.

Chickpeas contain higher levels of saponin than quinoa, while spinach, asparagus, parsley, garlic, onion, and many legumes contain lower levels. Possible saponin bitterness is the major problem with consuming quinoa grains. In terms of total saponins, values in the region of 0.9 [78] to 1.4 g 100 $g^{-1}$ [80] have been reported for bitter types. Quinoa containing 0.11 g 100 $g^{-1}$ of saponin or less can be considered sweet [69]. Quinoa seeds contain trypsin inhibitors in the range of 1.4 to 5.0 TIU (trypsin inhibitory units $mg^{-1}$ of seeds), but trypsin inhibitory activity may be absent [78]. These values are considerably lower in comparison to beans (12.9 to 14.8 TIU $mg^{-1}$ of seeds) and soya, however (24.5 TIU $mg^{-1}$ of seeds), and these inhibitors can be inactivated through the process of cooking. Soya seeds also contain phytic acid, which forms insoluble complexes with multivalent cations ($Ca^{2+}$, $Fe^{2+}$, $Fe^{3+}$ in $Zn^{2+}$) and consequently reduces their biological value. Present substances are also found in other types of food. In order not to present major hindrances to the use of quinoa, popular opinion asserts that it is possible to reduce or even eliminate these substances by selection and still preserve the high content of amino acids.

Quinoa is suitable for people with celiac disease because it does not contain gluten.

### 3.2.7   Gastronomy and suggestions for homemade food

Quinoa is very useful in a wide variety of foods. Traditionally, quinoa flour has been used for coarse-grained bread called *krispina* [81] and can also be fermented to make a beer called *chicha* [82]. Based on these traditions, the whole grains are used for soups, stews, and broths. Whole grains are also suitable for replacing rice-like products, and the coarse fractions of milling can be made into porridge, flakes, and puffed grains. Today, quinoa flour is often added to wheat flour to prepare food like breads, tortillas, cookies, salads, and pancakes. Many culinary ideas for preparing dishes based on quinoa are available on the Web.

### 3.2.8   Processing

Food processing from quinoa grains is developed and described [49], but for declared organic products, adaptations should be made according to organic food processing standards. The first step in processing is the washing or abrasion of pericarp, or their combination, due to the removal of saponins from the grains or flour. The combination of dry abrasive polishing and washing seems to be most effective, because losses of nutritive substances are minimized [77, 78].

Quinoa grain is often milled whole due to its very small grain size. Whole grain flour may include some or no pericarp if the pericarp was removed by abrasion, or it may include pericarp if the saponins from grains were washed before milling. Using the roller mill makes it possible to reduce quinoa grains into bran and flour fraction. The bran fraction is very rich in protein (usually more than 20%), because the bran consists mainly of embryo [50].

An industrial approach based on gun puffing, extrusion cooking, and expanded snack-type products [69] like flafes and poppies (Figure 3.2.1) is appropriate for organic food processing. This approach also makes it possible to prepare quinoa porridge flour [83], gluten-free pasta [84], or malted grain flour called "power" flour [85]. Quinoa instant infant porridge meets the energy requirements for children and contains about 16% proteins with a 95% rich digestibility rate [69].

Quinoa flour has been used to make leavened bread only as a 5% addition to wheat flour, because quinoa does not contain gluten. If a higher percentage of quinoa flour is added, the effect of the small-size starches and their amylase activity will result in very poor loaf volume. Good experiences are reported when preparing cakes [86] and cookies [87]. Cookies with an additional 20% of quinoa flour and cakes with up to 10% of quinoa flour have a delicious nutty taste.

## 3.3  Grain amaranths

### 3.3.1  Introduction

Since the 1970s, grain amaranths have been attracting increased attention due to the importance of their rich nutritional compounds to human nutrition. Grain amaranths are a small group from the genus *Amaranthus*, with more than 60 existing species. Plants are cosmopolitan, with extreme genetic diversity consisting of mostly weed amaranth species. The center of this biodiversity is central South America, containing numerous species, where they began with the cultivation of some species (*Amaranthus cruentus* L., *A. hypochondriacus* L., *A. caudatus* L.). Amaranth was one of the five irreplaceable food sources of the Aztecs in pre-Colombian Central American civilization. They grew it mostly in the Anahuac valley, but the colonization period reduced its production.

Amaranth, or huautli, also carried religious and agronomical importance besides its nutritional value. From water and amaranth, the atole drink was produced, grain was milled into flour (uauhatolli), and goudh from the flour was filled with leaves and prepared as an amaranth plate (huauquillamal-maliztli, translated as: a dish of bledos, prince's father tamales). For religious purposes, special figurines were made out of flour and honey. The shapes and sizes of the figurines depended on monthly rituals and varied from small pyramids to large images of mountain divinities. Those idols were carried around, and pieces of them were eventually eaten. To the colonizers, these rituals resembled Christian Eucharistic ceremonies; accordingly, they halted the amaranth production and banned its consumption. Amaranth was replaced by other cereals from the Old World, poverty became viewed as a negative status symbol, and amaranth's taste and religious connotations began to cause disapproval [88*, 89**].

Above all others, *A. caudatus* L. is considered the primary crop among South American civilizations. In the precolonization, period it was produced in Equador (where it was known as ataco and sangoracha), Bolivia (coimi, millmi), Peru (achiata, achos, achis, incajataco, coimi, kiwicha) [90], and Argentina. Amaranth's moisture was most significant to the Andes farmers. Amaranth is also used as an ornamental plant, a vegetable, in medicine, for feed use, and as a component in the cosmetic industry [88].

Commercial cultivation of amaranth for human nutrition does take place. In many developed countries, including the U.S. and some European countries, the amaranth's use has been extendedmainly to rich nutritional food produced under ecological–organic production systems.

## 3.3.2 Botany of grain amaranths

Amaranth is a pseudocereal, assigned to the family of dicotyledonous *Amaranthaceae*. Many species from genus *Amaranthus* should be used like a grain pseudocereal; according to grain potential and breeding programs, the following three species are the leading ones:

- Bush greens, red amaranth, huautli (*Amaranthus cruentus* L., sin. *A. paniculatus* L.);
- Prince's feather (*Amaranthus hypochondriacus* L.);
- Love-lies-bleeding, Inca wheat, cat-tail, tumbleweed, Omca wheat, Quihuicha, tassel flower (*Amaranthus caudatus* L.).

Amaranth is an annual plant. The height of plants differs among the species and according to the ecological circumstances and production systems. The plants range in height from a few centimeters to 3.5 (4) meters. They have a taproot and well-developed lateral roots, and leaves are elliptical or egg-shaped and notched at the tip of the leaf blade. Inflorescence levels differ among species. In general, the amaranth seeds are relatively light and small

* Reprinted from Grain amaranth, in *Cereals and Pseudocereals*, Williams, J.T., Brenner, D., Chapman and Hall, London, 1995, Copyright 2006, with permission from Kluwer Academic Publishers.
** Reprinted from Cultivated grain amaranths, in *Nekatere zapostavljene in/ali nove poljščine (Some disregarded and/or new field crops)*, Bavec, F., Univerza v Mariboru, Fakulteta za kmetijstvo, Maribor, 2000, 141.

with diameters of about 1 mm. With regard to *A. hypochondriacus,* inflorescence is large and branched and of a unified green or red color. Its numerous flowers contain sharp and rough bracts. The seed coat of amaranth is smooth and thin; thus, the seeds can be used directly in most cases.

The structure of the mature seed consists of a peripheric embryo surrounding the nutritive tissue, which is called a perisperm. Endosperm remnants are close to the root tip. The embryo encloses the starch-rich perisperm like a ring, and the percentage of embryo weight is about 25% of the grain weight. Inside the embryo a differentiation of the three primary meristematic tissues appears: the procambium appears as a single bundle in the embryonic axis or as small bundles throughout the cotyledon's length; these provascular cells are small and elongated and with fewer reserves and more cellular organelles than the large protoderm and ground meristem cells. These latter cells have more protein bodies, and they show a higher number of larger globoid crystal inclusions than the others. The perisperm is a starchy tissue, and its cells have thin walls and are full of angular starch grains [91*].

The plants of *A. cruentus* grow up to 180 cm tall. Inflorescence is green or red, sometimes spotty and soft, with bracts gentle that are to the touch (Figure 3.6). Seeds appear in various colors. The same can be said about the colors of inflorescence, which, in most cases, appear to be straighter than *A. hypochondriacus.*

*A. caudatus,* originally from Andes, is a plant that grows between 0.4 and 3.0 m tall. Leaves are attached with petioles, apposite or alternate, of a green to purple violet color. Inflorescence (Figure 3.6) can be classified from straight to horizontal, with lively colors including green, yellow, orange, pink, red, violet, and brown. Flowers are small. Male flowers consist of three to five stamina, and female flowers contain a superior ovary with one seed. Seeds are very small, only growing from 1 to 1.5 mm in diameter, and are mostly white (sometimes yellowish, reddish or black) and smooth. The 1000 to 3000 grain weight equals 1 g. In general, plants intended for grain production have lighter grain, whereas the grain of plants intended for vegetable production is darker. The level of allogamy in flowers is 10 to 50%. Cross-fertilization depends on wind, insects, pollen production, and so forth, and seeds mostly do not undergo dormancy.

In all analyzed characters (seed yield, stand density and height, and 1000 seed weight), there were more marked differences between *A. hypochondriacus* varieties and *A. cruentus* varieties [92].

## 3.3.3   Ecology

Amaranths perform photosynthesis via $C_4$ pathway. This characteristic allows for the responsible and efficient use of low concentrations of carbon dioxide by fixing in the chloroplasts of specialized cells surrounding the leaf

---

* Reprinted from *Annals. of Botany,* 74 (4), Coimbra, S. and Salema, R., *Amaranthus hypochondriacus*: Seed Structure and Localization of Seed Reserves, 373, Copyright 2005, with permission of Oxford University Press.

*Figure 3.6* Inflorescence of *A. caudatus.*

vascular bundles, and also results in low losses by transpiration in association with stomata and osmotic adjustment. This characteristic enables amaranths to achieve active photosynthesis at high temperatures with a tolerance to lack of water.

In general, short day illumination is most suitable for *A. caudatus*, although it is well adapted to all growth conditions. It flowers even with 12 to 16 hours of day illumination [88]. In all grain amaranth species, a light regime with 12 hours of illumination gave the highest percentage of live seed emergence [93].

Water demands vary from 400 to 800 mm, and  acceptable production is possible with only 250 mm of precipitation. Amaranth is a potentially useful crop in regions where soil moisture conditions vary considerably among growing seasons, due to its apparent ability to respond to water stress by increasing rooting depth. Maximum effective rooting depth of soil water extraction varies from 122 cm in less water-stressed years to 154 cm in more stressed years, occurring at early to late anthesis stages [94]. Highest amounts of water are required at the stage of seed emergence, germination, and flowering; during other stages, formed plants are tolerant to drought. Plants also adapt to moisture with more than 1000 mm of precipitation annually. Under dry conditions, approximately 70 to 75% of total water use (267 mm) occurred by the end of anthesis [94]. However, the year x genotype interaction is often significant for water use efficiency, plant height, plant aboveground biomass, grain yield, and harvest index.

The percentage of live seed emergence was above 80% in temperatures that exceeded 21°C [93]; in hybrid *A. hypochondriacus* x *A. hybridus*, the same percentage was achieved in temperatures that only exceeded 16°C [95]. The germination speed was closely related to temperature and decreased with decreasing temperature. Young plants are sensitive to warm temperature; during the branching stage, they can tolerate only up to 4°C. Unusually cool years and late sowing dates can result in plants not maturing before the first frost [96]. The maximum temperature for growth and development falls between 35 and 40°C.

## 3.3.4   Disposition for organic cultural practice

### 3.3.4.1   Growth patterns

Amaranth grows well in light sandy soils with high nutrient content. Soil should be airy or well drained. Suitable pH is from 6 to 7, but it also grows in extremely alkaline or acidic soil, which indicates tolerance to aluminum toxicity.

Regardless of the source of genetic material and its adaptation (South America, Central America, Russia, and Asia), the growth period from sowing to technological maturity lasts an average of 4 months; the wider range is from 105 to 160 days, which can be compared to the growth period of corn. However, an understanding of crop response to planting date is essential when evaluating a potential new crop like grain amaranths. Generally, the 48 accessions of *Amaranthus caudatus* L., *A. cruentus* L., and *A. hypochondriacus* L. in the Brazilian Savannah flowered 45 days after emergence, and the plants were harvested after 90 to 100 days [97]. The *A. cruentus* BRS Alegria, originated from mass selection in the variety AM 5189 of the U.S., gave in double-cropping — after soybeans — an average yield of 2359 kg grain ha$^{-1}$ and 5650 kg ha$^{-1}$ for total biomass, in 90 days from emergence to maturity [98].

### 3.3.4.2   Cultivation systems

According to the traditional production praxis of Central and South American countries, there are many acceptable production techniques that can be used in organic agriculture. Acceptable techniques include the following: direct sowing, sowing into classic seedbeds, the planting of seedlings, intercropped sowing with corn, sowing in bands separated from other crops, sowing at the edges of fields, and sowing as ornamental plants or as an extensive crop [99]. Using monoculture practice, which ignores the traditional agricultural practice, is not allowed in organic agriculture. For an organic agriculture practice, we suggest the inclusion of amaranths in the crop rotation every 3 to 5 years.

On a proper seedbed or in case of direct seeding, we need 4 to 6 kg of selected seed ha$^{-1}$. Row spacing should be around 0.8 m; Jamriska [92] found that the stands with narrower row spacing (12.5 cm) produced, on the average, higher yields than the stands with wider row spacing (0.5 m). According to our own experiences, the target population in Slovenia is 40 plants m$^{-2}$. We usually recommend sowing from 150 to 200 seeds m$^{-2}$ and manual thinning after emergence to achieve a final density of 29 to 52 plants m$^{-2}$.

The main effect of row spacing on grain yield was not significant, but the interaction of row spacing, plant population, and environment indicated population yield ranking differences at the 30-cm row spacing but not at the 76-cm row spacing [96]. Considering the analyses of yield, plant mortality, and potential harvest difficulties, the moderate population (173000 plants ha$^{-1}$), 76-cm row spacing, and generally higher-yielding *A. cruentus* cultivars would be recommended above 30-cm row spacing and populations of 74000 and 272000 plants ha$^{-1}$ [97].

Seed storage lasting more than 1 year decreased the percentage of germination. An early harvest of homogenous and dense amaranth crops is recommended for amaranth seed [95]. According to Bavec and Grobelnik [93*], a sowing depth of 15 mm gave the highest seedling weight on sand and the highest percentage of live seed emergence on loam with all investigated species (*Amaranthus mantegazzianus, A. hypochondriacus, A. cruentus,* and *A. caudatus*). On the sandy loam, the percentage of live seed emergence was not affected by sowing depth up to 15 mm. *A. caudatus* gave the highest percentage of live seed emergence, whereas *A. cruentus* gave the lowest percentage but the heaviest seedlings. Percentage of live seed emergence was severely reduced on the loam where topsoil crusting occurred after a decrease in soil moisture content from field capacity ranging from –60 to –70 kPa, but not on the silt loam even when dry conditions were maintained throughout the experiment. At 0.5-cm sowing depth, high moisture (10 to 20 kPa) did not influence emergence [100].

Amaranth crop provision is similar to corn: all nutrients should be easily available. Organic fertilizers, plowing under, or harvesting remains are necessary for sufficient nutrient supply. Although amaranth yield is responsive to available nitrogen in the soil, a high level of available nitrogen can negatively affect grain harvest in terms of excessive plant height, increased lodging, and delayed crop maturity [101].

Weeding needs to be performed at least twice: for the interrow space of emerged plants and when the plants are 15 to 25 cm high. In cases of natural land dominated by *Pennisetum purpureum* L., weeding performed two to four times is optimal for growth and development of grain amaranths. Weeding two times at 2 and 5 weeks after transplanting is optimal for growth parameters (plant height, leaf number, and shoot fresh weight); weeding three times at 2, 5, and 8 weeks after transplanting is optimal for shoot dry weight

---

* Reprinted from *Eur. J. Agron.*, 17 (2), Bavec, F. and Grobelnik-Mlakar, S., Effects of soil and climatic conditions on emergence of grain amaranths, 93, Copyright 2005, with permission from Elsevier.

and grain yield; and weeding four times at 2, 5, 8, and 11 weeks after transplanting is optimal for grain protein content [102*].

## 3.3.5   Harvesting and yielding

Harvesting takes place before the plant reaches full maturity. The appropriate harvesting period is indicated by the yellowing of bottom leaves and dry seed. In traditional amaranth production, manually harvested plants cut 20 cm tall are tied into 15 to 20 plant sheaves. They are dried for a couple of weeks until leaves and stems are completely dry. Threshing used to be described as traditional "dancing" in some civilizations; people walked on dry sheaves and later cleaned the product by blowing into their open palms filled with seeds. Threshing may also be performed mechanically by seed-cleaning machines. In literature on the subject, we find optimistic predictions regarding the adjustment of cereal combines for the harvesting of amaranth due to modern technical advances.

Estimations of amaranth yield vary. Data on yields from the U.S. range from 800 to 1500 kg seed ha$^{-1}$; in Peru, from 2000 to 5000 kg seed ha$^{-1}$; and in Bolivia, from 900 to 4000 kg seed ha$^{-1}$. Green matter yield of American cultivars in southern Germany reached 1300 g m$^{-2}$ with cereal index below 30, and seed loss with manual harvesting was approximately 30% [103]. Studies of 15 cultivars of different species (*A. cruentus, A. hypochondriacus, A. hybridus*) of various origins grown in central European conditions showed yields from 1 to 270 g of seed dry matter m$^{-2}$ and from 290 to 1440 g m$^{-2}$ aboveground dry seed matter [104].

Storage facilities must be dry and airy. In proper conditions, seeds can remain in storage for more than 10 years. The data for transportation and storage recommends a moisture level below 14% in the seeds. Data described by Abalone et al. [105] found that when moisture content changed from 7.7 to 43.9% dry base, true density decreased from 1390 to 1320 kg m$^3$, bulk density from 840 to 720 kg m$^3$, specific volume increased from 0.78 to 1.10 × 10$^3$ m$^3$ kg$^1$, and porosity ranged from 0.40 to 0.45.

## 3.3.6   Nutritional value

It is difficult to classify amaranth as food. According to general nutritional divisiona of food into carbohydrates (cereals, tubers), proteins (legumes and other nutrient sources), and minerals and vitamins (fruits and vegetables), amaranth leaves can be considered a vegetable, its grain is a rich source of carbohydrates, and its protein content is 12 to 16% (according to U.S. data). Lysine, methionine, cisteine, and arginine were found in the wholemeal of amaranth, while corn, barley, and common wheat proved to be poor in methionine and lysine [106]. Chemical scores of essential amino acids and essential amino acid in grain amaranths show the favorable nutritional

---

* Reprinted from *Crop Prot.*, 16 (5), Ojo, D.O., Effect of weeding frequencies on grain amaranth (*Amaranths cruenthus* L.) growth and yield, 463, Copyright 2005, with permission from Elsevier.

quality of amaranth protein, which is almost comparable with egg protein. The relatively high content of essential amino acids in amaranth grain pre-determines its use as a substitution of meat-and-bone meals [107].

The poor methionine and lysine content of common wheat could be supplemented mostly by adding the wholemeal of amaranth [106]. In 15 cultivars of prevailing species in central Europe, protein content of 13.3 to 17.9% in seeds was achieved. Muchova et al. [108*] compared all grain amaranth species and determined the highest content of the crude protein for *A. cruentus* (17.2% on average), whereas the lowest was *A. hypochondriacus* (15.72% on average). The starch contents were 543 in *A. caudatus* and 623 g kg$^{-1}$ in *A. cruentus*, while crude protein contents were 154 and 169 g kg$^{-1}$, respectively [109**].

Grain amaranth has been suggested as an alternative natural source of oil and squalene. The oil contents of grain amaranth are low (from 5.1% to 7.7%) as compared to other oil-containing grains; high concentrations of squalene were found in total lipids, ranging from 3.6% to 6.1%. The major fatty acids in amaranth oil consist of palmitic acid (19.1–23.4%), oleic acid (18.7–38.9%), and linoleic acid (36.7–55.9%) [110***].

Differential milling of amaranth yields three granulometric fractions as follows: high-fiber fraction, high-protein fraction, and high-starch semolina. The high-fiber fraction contains 63.9% insoluble fiber and 6.86% soluble fiber [111].

Protein fraction distribution in milling and screened physical fractions showed that the 30 mesh sample contained 2.34 fat and 9.05% protein, while the 40 mesh contained 16.18% fat and 26.46% protein [112].

## 3.3.7 Food processing

Twelve groups of food from amaranths and 12 processes (cooking, puffing or popping, milling, milling in classification, cooking and flaking, drum drying, cooking extrusion, cooking extrusion and milling, cooking extrusion and flaking, germination – malting, direct starch hydrolysis, and starch iso-lation) are described by Berghofer and Schoenlechner [113], but some of them need to be adapted according to organic food guidelines and standards.

As a rule, the seeds are hard and the flower is very useful. Whole seeds can also be prepared. More than 50 ways of preparing amaranths for food are known. Leaves are still used in salads, soups, creams, desserts, drinks, and bread, and cakes are made out of seeds; additionally, organic cereals are now consumed in Austria. When making bread, 20% of flour is usually

* Muchova, Z., Cukova, L., and Mucha, R., Seed protein fractions of amaranth (*Amaranthus* sp.), *Rost. Vyroba*, 46 (7), 331, 2000.
** Reprinted from *J. Sci. Food Agric.*, 84 (10), Gamel, T.H., Nutritional study of raw and popped seed proteins of *Amaranthus caudatus* L and *Amaranthus cruentus* L., 1153, Copyright 2005, with permission from Elsevier.
*** Reprinted from *J. Agric. Food. Chem.*, 50 (2), He, H.P. et al., Extraction and purification of squalene from Amaranthus grain, 368, Copyright 2005, with permission from Elsevier.

substituted. Substitution of 10% and 15% (on weight base) wheat flour by amaranth flour has a positive effect on dough quality (increased binding of flour, better dough processing), the amount of produced $CO_2$ (increased), porosity of bread inside (more regular with softer pores), and nutritive value of products (increased). A considerably decreased gluten content and a negative effect on dough quality (adhesiveness) and bread (very low specific volume, considerable amaranth flavor) for 20% (on weight base) substitution of wheat flour by amaranth flour was found by Burisova et al. [114].

Grain amaranth has unique microcrystalline starch granules (1 to 3 mm in diameter). Pearled and unpearled *A. cruentus* seed was wet-milled using a high-alkaline, batch-steeping process and separation methods common to laboratory wet-milling of corn. Starch with a purity of 0.2% protein was obtained from both the pearled and unpearled amaranth. However, more starch was recovered from unpearled amaranth because of the leaching of fine starch granules during steeping. Less germ was recovered using unpearled amaranth [115*].

The share of the glutelins ranged (on average) from 15.45% for *A. hypochondriacus* to 20.63% for *A. cruentus*. The observed differences between the pairs of the species were all highly statistically significant, except for the difference between *A. caudatus* and *A. cruentus*, which was only somewhat significant, and between *A. hypochondriacus* and *A. paniculatus*, which was not significant at all. In relation to the percentages of the nutritionally important protein fractions (i.e., albumins + globulins + insoluble remnants) to the nutritionally least important prolamins fraction, we can line the nutritional value of the studied species up in the following order: *A. paniculatus, A, caudatus, A. cruentus,* and *A. hypochondriacus* [108].

Several treatments, including cooking, popping, germination, and flour air classification, mainly affected the protein and starch properties [109]. Heat treatment by seed popping at 170 to 190°C for 30 seconds resulted in significantly ($P < 0.05$) decreased valine and leucine contents. High contents of lysine and arginin were detected in both heat-treated and untreated grains, as well as satisfactory content of cysteine and lower levels of methionine, valine, lysine, and leucine. The latter three amino acids appear as limiting [107]. After popping, the true protein content in *A. caudatus* and *A. cruentus* decreased by 9 and 13%, respectively. Among the amino acids, the loss of tyrosine due to the popping effect was the highest, followed by phenylalanine and methionine. Leucine was the first limiting amino acid in the raw samples, followed by lysine, while the reverse order was observed in the popped samples [109].

Seed and flour can be used for making chocolate powder, syrups, and cakes. The study conducted on using green parts for natural organic dyes did not provide promising results.

---

* Reprinted from *Cereal Chem.*, 71 (1), Myers, D.J. and Fox, S.R., Alkali wet-milling characteristics of pearled and unpearled amaranth seed, 96, Copyright 2005, with permission from the American Association of Cereal Chemists.

*Figure 3.7* Breakfast made from amaranth seeds.

Amaranth flour has been found suitable for pasta products with excellent cooking and sensory properties [116]. Results suggest that for at least the soaking temperatures above 64°C, the absorption process of water is controlled by water–starch reactivity [117]. Without emulsifier, the production of pasta products was not possible [116]; however, emulsifier should be permitted for organic food processing.

Amaranth (*Amaranthus caudatus* L.) extrusion produced a highly acceptable snack product based on amaranth grains and flour (Figure 3.7). The most expanded products also had the best textures and were obtained at 150°C and 15% moisture. These conditions resulted in greater expansion, greater shearing force of extrudates, greater extrudate surface area per unit weight, and reduced shearing stress at maximum shearing force [118*]. A processing method for producing a high-protein flour and maltodextrins from whole amaranth flour was developed. The protein-enriched flour (31% of protein) may be used as a dry milk extender [119].

The new option in food processing might be a protein concentrate from amaranths described by Escudero et al. [120]. In cases when the flour protein content was 16.6 g% while that of the concentrate was 52.56 g%. According to the amino acid composition suggested from FAO indicated that the concentrate does not have limiting amino acids. The content of lysine was high in both the flour and the concentrate, making these products particularly useful as complements to cereal flour, which is deficient in this amino acid. The content of the soluble dietary fiber with a hypolipemic function was notably higher in the protein concentrate (12.90 g%) than in the seed flour (4.29 g%). The protein concentrate also exhibited a higher content of insoluble dietary fiber. The flour and the concentrate contain 75.44 and 56.95% of

* Reprinted from *J. Food Sci.*, 65 (6), Chavez-Jauregui, R.N., Silva M.E.M.P., and Areas, J.A.G., Extrusion cooking process for amaranth (*Amaranthus caudatus* L.), 1009, 2000, Copyright 2006, with permission from the Institute of Food Technologists.

unsaturated fatty acids, respectively. Squalene was detected both in the flour and the concentrate oils, with a higher content in the concentrate (9.53%) than the flour (6.23%). The presence of trypsin inhibitors, saponins, and phytic acid in the concentrate favor the metabolism of lipids; Escudero et al. [120] speculate that consumption of the concentrate might actually reduce the risk of heart disease. Also, combining the fractions rich in squalene gave a special 94% squalene concentrate [110].

### 3.3.8   Health value

Amaranths constitute an alternative source of proteins in the human diet, with advantages over animal proteins because of their low content of saturated fats and absence of cholesterol [120]. The amaranth grains, with excellent protein quality, can also be used in gluten-free special diets, especially for children.

Analyses proved that inclusion of amaranths in nutrition positively influenced hemoglobin or prevented blood anemia. The inclusion of amaranth in everyday diets can help avoid $FeSO_4$ diets due to 61% of biological acceptability of Fe from *A. hypochondriacus*. From this point of view, green leaves are also a quality vegetable [121].

The most important constituent of amaranth is squalene (affects biosynthesis of cholesterol), which, until the present time, was extracted from the livers of whales and sharks. Introduction of this plant to food products acts prophylactically, i.e., it appears to possess antisclerotic properties and combat constipation. This latter property may be used in the prophylaxis against cancer [122].

# 3.4.   Wild rice

## 3.4.1   Introduction

Wild rice, also called indian rice and canadian rice (*Zizania palustris* L. and *Z. aquatiaca* L.), was a staple food for indigenous North Americans and arriving Europeans for thousands of years. It has also been cultivated in temperate regions of Asia. Interest in rice production is growing all over the world; production possibilities have been researched by both the Japanese [123] and Finns [124]. The plant is an aquatic grass that grows naturally in lakes and slowly flowing rivers, i.e., in moderately soft and acidic freshwater wetlands. However, for organic production, contaminated water without environmentally dangerous compounds is essential.

## 3.4.2   Biology

Within *Zizania* genus, modern taxonomists distinguish four wild rice species (*Zizania latifolia* [Griseb.] Turcz. & Stapf, *Z. texana* Hitchock, *Z. aquatica* L., and *Z. palustris* L.). *Zizania palustris* L. var. *palustris* (0.5 to 1.5 m high plants) and var. interior (0.9 to 3.0 m high plants) have the highest commercial value.

The florescence is a terminal panicle with male spikelets in the lower branches and female spikelets in the upper branches; the spikelets involve the formation of staminate and pistillate floral primordia. The frequency of plants with hermaphroditic floret formation ranged from 27% in the pistillate population (*Z. palustris*) to 70% in the Peterson Pond (*Z. aquatica*) population [125, 126].

### 3.4.3 Growth and cultivation

Production of wild rice has many specific characteristics, especially due to its biological, physiological, and postharvest possibilities. Its main characteristic is that the plant is produced on the paddies of the flooded fields with high organic content in the soils, with pH ranging from 5.8 to 7.8. The water depth in the rice paddies should vary between 0.2 and 0.6 m. The common seeding method in the new paddies involves broadcasting the seed at a rate of 30 to 50 kg viable seed ha$^{-1}$. In natural stands, the seeds mature in the autumn and pass the winter on the flooded paddy bed, where they reseed themselves. The highest yielding plant population does not exceed approximately 10 plants m$^{-2}$.

In natural stands, the seeds mature in the autumn and pass the winter on the lake bed. They germinate in the spring, with growth to maturity requiring approximately 100 days. Japanese research [123] found that seedling growth was much better in plants grown at 20°C than those at grown at 12°C, but there were no interactions between these temperatures and planting in water depths of 2, 6, and 8 cm, suggesting that temperature is an independent factor of water depth [123].

Because the plant is an aquatic grass, it grows in artificially flooded paddies, margins of lakes, and slowly flowing rivers and brooks; in the long term, the balance of nutrients may be affected by straw after harvest and complete decomposition may require 3 years.

Harvest maturity is reached when approximately one-third of the seeds are dark-colored and possess about 40% moisture. To protect natural stands in the state of Minnesota since 1939, the law has provided that wild rice in public waters can be harvested only by hand [127]. Today, combines or other mechanical harvesters are used for harvesting. If an aquatic weed harvester were used to remove standing straw from part of a wild rice stand at the end of the growing season, no difference would be noted in rates of organic matter decomposition or nutrient release between the chopped and unchopped straw [128].

### 3.4.4 Nutritional value and utilization

The protein content of wild rice is between 12 and 15%. It is important that wild rice contain 3.8 to 4.2% of lysine within the entire protein content; as such, wild rice contains twice as much lysine as other lysine-deficient cereals. Even lysine maize hybrids contain less lysine than wild rice. Wild rice contains more iron and zinc than brown or white polished rice and other cereals, as well.

Antioxidants have been isolated from wild rice seeds [129]. The high concentrations of iron (2.0–9.7 mg 100 $g^{-1}$ dry weight), copper (0.2–1.3 mg 100 $g^{-1}$ dry weight), and zinc (0.1–0.4 mg 100 $g^{-1}$ dry weight) in 26 brands of wild rice suggest that wild rice may be a good dietary source of these essential elements [130].

The grain yield varies from just a few hundred kilograms to 1250 kg $ha^{-1}$ in commercial paddies. However, organically produced plants, especially in commercial wild rice production areas, can be attacked by several plant diseases causes by fungal brown spot (*Bipoolaris oryze* and *B. sorokiniana*), stem rot (*Helmintosporium sigmoidinum* and *Sclerotinium* sp.), leaf smut (*Entyloma lineatum*), ergots (*Claviceps purpurea*) [131], *Phythophthora erythroseptica* [132], and insect pest (*Apamea apamiformis*) [133].

Wild rice production is a promising approach in some environments to the reduction of pests and diseases. Common rice in the free market belongs to the food most often treated with pesticides due to monoculture production in export countries. In the U.S. and Canada, wild rice is often grown and sold as an organic food and processed product; many producer and seller associations have been established for this purpose. Organic wild rice is the food that has spread the farthest among organic shops throughout the world. Food is prepared from wild rice by cooking the grains in boiled and salted water (with possible variations in water levels) for 25 to 35 minutes, bringing it to a simmer, and leaving it covered for 5 minutes with the heat turned off. The boiling times may vary according to the desired. However, wild rice is extremely versatile, allowing every cook vast opportunities to feature his or her culinary talents. Preparing food based on sole or prevailing compounds of wild rice grains — such as soups, salads, sweets, and side dishes — is a special pleasure and can be very tasty. For this reason, the popularity of wild rice is steadily increasing. Many recipes for innovative cooks are currently available on the Web.

## References

1. Leenders, http://www.theorb.net/ encyclop/high/low_count/essays/text 06.html (accessed November 2004).
2. Kreft, I. et al., *Ethnobotany of Buckwheat*, Jinsol Publishing Co., Seul, 2003.
3. Scharm, D.D. et al., Honey with high levels of antioxidants can provide protection to healthy human subjects, *J. Agric. Food Chem.*, 51 (6), 1732, 2003.
4. Bavec, F., Ajda, buckwheat (*Fagopyrum esculentum*), in *Some of Disregarded and/ or New Field Crops* (Slovene language), University of Maribor, Faculty of Agriculture, Maribor, 2000, 6.
5. Murakami, T. et al., Root length and distribution of field-grown buckwheat (*Fagopyrum esculentum* Moench), *Soil Sci. Plant. Nutr.*, 48 (4), 609, 2002.

6. Peng, Y.Y., Liu, F.H., and Ye, J.N., Determination of phenolic compounds in the hull and flour of buckwheat (*Fagopyrum esculentum* Moench) by capilar electrophoresis with electrochemical detection, *Analytical Letters*, 37 (13), 2789, 2004.

7. Bavec, F., Pušnik, S., and Rajčan, I., Yield performance of two buckwheat genotypes grown as a full-season and stubble-crop, *Rostl. Vyroba*, 48 (8), 351, 2002.

8. Semwal, R.L. et al., Patterns and ecological implications of agricultural land-use changes: A case study from central Himalaya, India, *Agric., Ecosyst. Environ.*, 102 (1), 81, 2004.

9. Kalinova, J. and Moudry, J., Evaluation of frost resistance in varieties of common buckwheat (*Fagopyrum esculentum* Moench), *Plant Soil and Environ.*, 49 (9), 410, 2003.

10. Sugimoto, H. and Sato, T., Effects of excessive soil moisture at different growth stages on seed yield of summer buckwheat, *Japanese J. Crop Sci.*, 69 (2), 189, 2000.

11. Edwardson, S.E., Using growing degree days to estimate optimum windrowing time in buckwheat, in *Proc. Int. Symp. Buckwheat*, Japan, Shinshu Univ. Nagano, II, 26, 1995.

12. Slawinska, J. and Obendorf, R.L., Buckwheat seed set in planta and during in vitro inflorescence culture: Evaluation of temperature and water deficit stress, *Seed Sci. Res.*, 11 (3), 223, 2001.

13. Delperee, C., Kinet, J.M., and Lutts, S., Low irradiance modifies the effect of water stress on survival and growth-related parameters during the early developmental stages of buckwheat (*Fagopyrum esculentum*), *Physiol. Plantarum*, 119 (29), 211, 2003.

14. Michiyama, H. et al., Influence of day length before and after the start of anthesis on the growth, flowering and seed-setting in common buckwheat (*Fagopyrum esculentum* Moench), *Plant Prod. Sci.*, 6 (4), 235, 2003.

15. Aufhammer, W. and Esswein, H., Productivity of buckwheat (*Fagopyrum esculentum*) as an alternative crop, in *Proc. 2nd ESA Congr.*, Warwick University, Scaife, A., Ed., ESA UK Congress Office, Warwick, 1992, 28.

16. Kreft, I., *Ajda (Buckwheat)*, ZP Kmeki glas, Ljubljana, 1995.

17. Aufhammer, W., Esswein, H., and Kubler, E., Zur entwicklung und nutzbarkeit des körnerertragspotentials von buchweizen (*Fagopyrum esculentum*), *Journal für Landwirtschaftliche Forschung*, 45 (1), 37, 1994.

18. Tsuzuki, E. and Dong, Y.J., Buckwheat allelopathy: Use in weed management, *Allelopathy J.*, 12 (1), 1, 2003.

19. Iqbal, Z. et al., Allelopathic activity of buckwheat: Isolation and characterization of phenolics, *Weed Sci.*, 51 (5), 657, 2003.

20. Bavec, M. et al., Buckwheat leaf area index and yield performance depending on plant population under full-season and stubble-crop growing periods, *Die Bodenkultur, in press.*

21. Goodman, R. et al., Honeybee pollination of buckwheat (*Fagopyrum esculentum* Moench) cv. Manor, *Australian J. of Exp. Agric.*, 41 (8), 1217, 2001.

22. Shonbeck, M. et al., Comparison of weed biomass and flora in 4 cover crops and subsequent lettuce crop on 3 New England organic farms, *Biol. Agric. Holtic.*, 8 (2), 123, 1991.

23. Zhu, Y.G. et al., Buckwheat (*Fagopyrum esculentum* Moench) has high capacity to take up phosphorus (P) from a calcium (Ca)-bound source, *Plant Soil.*, 239 (1), 1, 2002.
24. Edwardson, S., Buckwheat: Pseudocereal and nutraceutical, in *Progress in New Crops*, Janick, J., Ed., ASHS Press, Alexandria, VA, 1996, 195.
25. Pomeranz, Y., Buckwheat structure, composition, and utilization, *Crit. Rev. Food. Sci. Nutr.*, 19 (3), 213, 1981.
26. Robinson, R.G., *The Buckwheat Crop in Minnesota*, *Agr. Exp. Sta. Bul.*, Univ. Minnesota, St. Paul, 1980.
27. Belton, P. and Taylor, J., *Pseudocereals and Less Common Cereals. Grain Properties and Potential*, Springer-Verlag, Berlin, 2002.
28. Steadman, K.J., Minerals, phytic acid, tannin and rutin in buckwheat seed milling fractions, *J. Sci. Food Agric.*, 81 (11), 1094, 2001.
29. Thacker, P.A., Anderson, D.M., and Bowland, J.P., Buckwheat as a potential feed ingredient for use in pig diets, *Pig News Inform.*, 5 (2), 77, 1984.
30. Kreft, I. et al., New nutritional aspects of buckwheat based products, *Getreide-Mehl und Brot*, 52, 27, 1998.
31. Wei, Y.M. et al., Studies on the amino acid and mineral content of buckwheat protein fractions, *Nahrung-Food*, 47 (2), 114, 2003.
32. Krkoškova, B. and Mrazova, Z., Prophylactic components of buckwheat, *Food Res. Int.*, 38 (5), 561, 2005.
33. Li, S.Q. and Zhang, Q.H., Advances in the development of functional foods from buckwheat, *Crit. Rev. Food Sci. Nutr.*, 41 (6), 451, 2001.
34. De Francischi, M.L., Salgado, J.M., and Da Costa, C.P., Immunological analysis of serum for buckwheat fed celiac patients, *Plant. Foods Hum. Nutr.*, 46, 207, 1994.
35. Watanabe, M., Catachins as antioxidants form buckwheat (*Fagopyrum esculentum* Moench) groats, *J. Agric. Food Chem.*, 46, 839, 1998.
36. Pisha, E. and Pezzuto, J.M., Fruits and vegetables contain compounds that demonstrate pharmalogical activity on humans, in *Economic and Medical Plant Research*, Wagner, H., Hikino, H., and Farnswoth, N.R., Eds., UK Academic Press, London, 1994, 189.
37. Dietrich-Szostak, D. and Oleszek, W., Effect of processing on the flavonoid in buckwheat (*Fagopyrum esculentum* Moench) grain, *J. Agric. Food Chem.*, 47, 4384, 1999.
38. Jiang, H.M.J. et al., Oats and buckwheat intakes and cardiovascular disease risk factors ain an ethnic minority in China, *Am. J. Clin. Nutr.*, 61, 366, 1995.
39. Udesky, J. and Sturtleff, W., *The Book of Soba*, Harper & Row, New York, 1995.
40. Tomatoke, H., A buckwheat protein product suppresses gallstone formation and plasma cholesterol more strongly than soy protein isolate in hamsters, *J. Nutr.*, 130, 1670, 2000.
41. He, J. et al., Oats and buckwheat intakes and cardiovascular disease risk factors in an ethnic minority of China, *Am. J. Clin. Nutr.*, 61, 366, 1995.
42. Prestamo, G. et al., Role of buckwheat diet on rats as prebiotic and healthy food, *Nutr. Res.*, 23 (6), 803, 2003.
43. Deschner, E., Dietary quercetin and rutin: Inhibitors of experimental colonic neoplasia, in *Phenolic Compounds in Food and Their Effects on Health II: Antioxidants and Cancer Prevention*, Huang, M.T., Ho, C.T., and Lee, C.Y., Eds., American Chemical Association, Washington, DC, 1992, 265.

44. Liu, Z. et al., Buckwheat protein product suppresses 1,2 dimethylhydrazine-induced colon carcinogenesis in rats by reducing cell proliferation, *J. Nutr.*, 131, 1850, 2001.
45. Noma, T. et al., Fatal buckwheat dependent exercised-induced anaphylaxis, *Asian Pac. J. Allergy Immunol.*, 19 (4), 283, 2001.
46. Datua, G. et al., Flour and allergy: Pitfalls which must be recognized, *Revue Francaise d allergologie et d immunologie clinique*, 42 (3), 289, 2002.
47. Kalinova, J., Moudry, J., and Curn, V., Technological quality of common buckwheat (*Fagopyrum esculentum* Moench), *Rostl. Vyroba*, 48 (6), 279, 2002.
48. Im, J.S., Huff, H.E., and Hsieh, F.H., Effects of processing conditions on the physical and chemical properties of buckwheat grit cakes, *J. Agric. Food Chem.*, 5 (3), 659, 2003.
49. http://www.pipedreamdesign.co.uk/quinoa/index.htm (accessed April 2006).
50. Bavec, F., *Chenopodium quinoa* Willd., in *Nekatere Zapostavljene in/ali Nove Poljšcine (Some of Disregarded and/or New Field Crops)*, Univerza v Mariboru, Fakulteta za kmetijstvo, Maribor, 2000, 167.
51. Mujica, A. et al., Quinoa (*Chenopodium quinoa* Willd.): Ancestral cultivo Andino, alimento del presente y futuro, project FAO presentation, Santiago, Chile, 2001.
52. Jacobsen, S.E. and Stølen, O., Quinoa — morphology, phenology and prospects for its production as a new crop in Europe, *Eur. J. Agron.*, 2, 19, 1993.
53. Prego, I., Maldano, S., and Otegui, M., Seed structure and localization of reserves *Chenopodium quinoa*, *Annal. Bot.*, 82 (4), 481, 1998.
54. Flores, F.G., Estudio preliminar de la fenologia de la quinoa (*Chenopodium quinoa* Willd.), Ingeniero Agronomo Thesis, Universidad Nacional Tecnica del Antiplano, PUNP– Peru, 1977.
55. Mujica, A., Andean grains and legumes, in *Neglected crops 1492 from a different perspective*, Hernández Bermejo, J.E., and León, J., Eds., FAO, Rome, 1994, 131.
56. Risi, C. Jr. and Galwey, N.W., The pattern of genetic diversity in the Andean grain crop quinoa (*Chenopodium quinoa* Willd.), I. Associations between characteristics, *Euphythica*, 41 (1–2), 147, 1989.
57. Alvarez, M. and Rütte, S., Fertilización, in quinoa: Hacia su cultivo commercial, Wahli, C.H. and Latinreco, S.A., Ed., Cassila 17-110-6053, Quito, Ecuador, 95.
58. Garcia, M., Raes, D., and Jacobsen, S.E., Evapotranspiration analysis and irrigation requirements of quinoa (*Chenopodium quinoa*) in the Bolivian highlands, *Agric. Water Manag.*, 60 (2), 119, 2003.
59. Milovanovič, M., The effects of changing dry periods on growth of quinoa (*Chenopodium quinoa* Willd.), Thesis, University of Maribor, Faculty of Agriculture, Maribor, 2004.
60. Jensen, C.R. et al., Leaf gas exchange and water relation characteristics of field quinoa during soil drying, *Eur. J. Agron.*, 13 (1), 11, 2000.
61. www.hort.pordue.edu/newcrop/afcm/quinoa.html (accessed March 2004).
62. Rasmussen, C., Lagnaoui, A., and Esbjerg, P., Advances in the knowledge of quinoa pests, *Food Reviews Int.*, 19, 61, 2003.
63. www-u.life.uiuc.edu/~ clausur/quinoa.html (accessed January 2005).
64. Schlick, G. and Bubenheim, D.L., Quinoa: Candidate crop for NASA's controlled ecological life support system, in *Progress in New Crops*, Janick, J., Ed., ASHS Press, Arlington, VA, 1996, 632.

65. Erley, G.S.A. et al., Yield and nitrogen utilization efficiency of the pseudoce-reales amaranth, quinoa, and buckwheat under differing nitrogen fertiliza-tion, *Eur. J. Agron.*, 22 (1), 95, 2005.

66. Johnson, D.L. and Ward, S., Quinoa, in *New Crops*, Janick, J. and Simon, J.E., Eds, Wiley, New York, 1996, 222.

67. Jacobsen, S.E., Jorgensen, I., and Stolen, O., Cultivation of quinoa (*Chenopo-dium quinoa*) under temperate climatic conditions in Denmark, *J. Agric. Sci.*, 122, 47, 2004.

68. Coulter, L. and Lorenz, K., Quinoa-composition, nutritional value, food ap-plications, *Lebens.-Wiess.+Technol.*, 23, 203, 1990.

69. Taylor, J.R.N. and Parker, M.L., Quinoa, in *Pseudocereals and Less Common Cereals: Grain Properties and Utilization Potential*, Belton, P. and Taylor, J., Eds., Springer-Verlag, Berlin, 2002, 93.

70. Galwey, N.W. et al., Chemical composition and nutritional characteristics of quinoa, *Food Sci. Nutr.*, 42, 245, 1990.

71. European Patent Application, No. 891216554.1, 1989.

72. Fleming, J.E. and Galwey, N.W., Quinoa (*Chenopodium quinoa*), in *Underuti-lized Crops: Cereals and Pseudocereals*, Williams, J.T., Ed., Chapman & Hall, London, 1995, 3.

73. Koziol, M.J., Chemical composition and nutritional evaluation of quinoa (*Che-nopodium quinoa* Willd.), *J. Food Composition Anal.*, 5, 35, 1992.

74. Ruales, J., Development of an infant food from quinoa (*Chenopodium quinoa*, Willd.): Technological aspects and nutritional consequences, PhD Thesis, Uni-versity of Lund, Lund, Sweden, 1992.

75. Telleria, M., Sgrabieri, V.C., and Amaya, J., Evaluación quimica y biológica de la quinoa (*Chenopodium quinoa* Willd). Influencia de la extracción de las saponinas por tratamiento térmico, *Arch. Latinoamer. Nutr.*, 28, 253, 1978.

76. Becker, R. and Hanners, G.D., Compositional and nutritional evaluation of quinoa whole grain flour and mill fractions, *Lebens-Wiess.+Technol.*, 23, 441, 1990.

77. Chauhan, G.S., Eskin, N.A.M., and Thachuk, R., Nutrients and antinutrients in quinoa seed, *Cereal Chem.*, 69, 85, 1992.

78. Ruales, J. and Nair, B.M., Content of fat, vitamins and minerals in quinoa (*Chenopodium quinoa* Willd) seeds, *Food Chem.*, 48, 131, 1993.

79. Guzmán-Maldonado, S.H. and Paredes-López, O., Functional products of plants indigenous to Latin America: Amaranth, quinoa, common beans, and botanicals, in *Functional Foods: Biochemical and Processing Aspect*, Mazza, G., Ed., Technomic Publishing, Lanchester, 1998, 293.

80. Gee, J.M. et al., Saponins of quinoa (*Chenopodium quinoa*): Effects of processing on their abudance in quinoa products and their biological effects on intestinal mucosal tissue, *J. Sci. Food Agric.*, 63, 201, 1993.

81. Weber, E.J., The Inca's ancient answer to food shortage, *Nature*, 272, 486, 1978.

82. Simmonds, N.W., The grain Chenopods of the tropical America highlands, *Econ. Bot.*, 19, 223, 1965.

83. Ruales, J., Valencia, S., and Nair, B., Effect of processing on the physical-chem-ical characteristics of quinoa flour (*Chenopodium quinoa* Willd), *Starch/Staerke*, 45, 13, 1993.

84. Caperuto, L.C., Amaya-Farfan, J., and Camargo, C.R.O., Performance of quinoa (*Chenopodium quinoa* Willd) flour in the manufacture of gluten free spaghetti, *J. Sci. Food Agric.*, 81, 95, 2000.

85. Mosha, A.C. and Svanberg, U., Preparation of weaning foods with high nutrient density using flour of germinated cereals, *Food Nutr. Bull.*, 5, 10, 1983.
86. Lorenz, K., Quinoa (*Chenopodium quinoa*) starch–physico-chemical properties and functional characteristics, *Starch/Staerke*, 42, 81, 1990.
87. Lorenz, K. and Coulter, L., Quinoa flour in baked products, *Plant Food Hum. Nutr.*, 41, 213, 1991.
88. Williams, J.T. and Brenner, D., Grain amaranth (*Amaranthus species*), in *Cereals and Pseudocereals*, Williams, J.T., Ed., Chapman and Hall, London, 1995.
89. Bavec, F., Cultivated grain amaranths, in *Nekatere zapostavljene in/ali nove poljščine (Some of disregarded and/or new field crops)*, Univerza v Mariboru, Fakulteta za kmetijstvo, Maribor, 2000, 141.
90. Sumar-Kalinovski, L., Kiwicha: El grano que se agiganta, *Ovonoticias*, 96 (9), 19, 1985.
91. Coimbra, S. and Salema, R., *Amaranthus hypochondriacus*: Seed structure and localization of seed reserves, *Ann. of Bot.*, 74 (4), 373, 1994.
92. Jamriska, P., The effect of variety and row spacing on seed yield of amaranth (*Amaranthus ssp.*), *Rostl. Vyroba*, 44 (2), 71, 1998.
93. Bavec, F. and Grobelnik-Mlakar, S., Effects of soil and climatic conditions on ermegence of grain amaranths, *Eur. J. Agron.*, 17 (2), 93, 2002.
94. Johnson, B.L. and Henderson, T.L., Water use patterns of grain amaranth in the northern Great Plains, *Agron. J.*, 94 (6), 1437, 2002.
95. Aufhammer, W. et al., Germination of grain amaranth (*Amaranthus hypochondriacus x A-hybridus*): Effects of seed quality, temperature, light and pesticides, *Eur. J. Agron.*, 8 (1–2), 127, 1998.
96. Henderson, T.L., Johnson, B.L., and Schneiter A.A., Grain amaranth seeding dates in the Northern Great Plains, *Agron. J.*, 90 (3), 339, 1998.
97. Henderson, T.L., Johnson, B.L., and Schneiter, A.A., Row spacing, plant population, and cultivar effects on grain amaranth in the northern Great Plains, *Agron. J.*, 92 (2), 329, 2000.
98. Teixeira, D.L., Spehar, C.R. and Souza L.A.C., Agronomic characterization of amaranth for cultivation in the Brazilian Savannah, *Pesquisa Agropecuaria Brasileira*, 38 (1), 45, 2003.
99. Spehar, C.R. et al., Amaranth BRS Alegria: Alternative for diversification of cropping systems, *Pesquisa Agropecuaria Brasileira*, 38 (5), 659, 2003.
100. Grobelnik-Mlakar, S. and Bavec, F., Environmental impact on emergence of seedling of Amaranth, presented at 4th Congr. of European Amaranth Association, Nitra, Slovakia, Aug. 16–19, 1999.
101. Myers, R.L., Nitrogen fertilizer effect on grain amaranth, *Agron. J.*, 90 (5), 597, 1998.
102. Ojo, D.O., Effect of weeding frequencies on grain amaranth (*Amaranths cruenthus* L.) growth and yield, *Crop Prot.*, 16 (5), 463, 1997.
103. Aufhammer, W. et al., Grain yield formation and nitrogen uptake of amaranth, *Eur. J. Agron.*, 4 (3), 379, 1995.
104. Kaul, H.P. et al., The suitability of amaranth genotypes for grain and fodder use in Central Europe, *Die Bodenkultur*, 47 (3), 173, 1996.
105. Abalone, R. et al., Some physical properties of amaranth seeds, *Biosyst. Eng.*, 89 (1), 109, 2004.
106. Matuz, J. et al., Structure and potential allergenic character of cereal proteins — I. Protein content and amino acid composition, *Cereal. Res. Commun.*, 28 (3), 263, 2000.

107. Pisarikova, B., Kracmar, S., and Herzig, I., Amino acid contents and biological value of protein in various amaranth species, *Czech. J. Anim. Sci.*, 50 (4), 169, 2005.
108. Muchova, Z., Cukova, L., and Mucha, R., Seed protein fractions of amaranth (*Amaranthus* sp.), *Rostl. Vyroba*, 46 (7), 331, 2000.
109. Gamel, T.H., Nutritional study of raw and popped seed proteins of *Amaranthus caudatus* L and *Amaranthus cruentus* L., *J. Sci. Food Agric.*, 84 (10), 1153, 2004.
110. He, H.P. et al., Extraction and purification of squalene from Amaranthus grain, *J. Agric. Food. Chem.*, 50 (2), 368, 2002.
111. Tosi, E.A., Dietary fiber obtained from amaranth (*Amaranthus cruentus*) grain by differential milling, *Food Chem.*, 73 (4), 441, 2001.
112. Bressani, M.E.B.S.Y.R., Protein fraction distribution in milling and screened physical fractions of Grain amaranth, *Alan.*, 52 (2), 167, 2002.
113. Berghofer, E. and Schoenlechner, R., Grain amaranth, in *Pseudocereals and Less Common Cereals. Grain Properties and Utilization Potential*, Belton, P. and Taylor, J., Eds., Springer-Verlag, Berlin, 2002.
114. Burisova, A. et al., The influence of substitution of wheat flour by amaranth flour on fermentative gas production and quality of bread, *Rostl. Vyroba*, 47 (6), 276, 2001.
115. Myers, D.J. and Fox, S.R., Alkali wet-milling characteristics of pearled and unpearled amaranth seed, *Cereal Chem.*, 71 (1), 96, 1994.
116. Kovacs, E.T., Maraz-Szabo, L., and Varga, J., Examination of the protein-emulsifier-carbohydrate interactions in amaranth based pasta products, *Acta aliment.*, 30 (2), 173, 2001.
117. Resio, A.N.C., Aguerre, R.J., and Suarez, C., Analysis of simultaneous water absorption and water-starch reaction during soaking of amaranth grain, *J. Food Eng.*, 68 (2), 265, 2005.
118. Chavez-Jauregui, R.N., Silva M.E.M.P., and Areas, J.A.G., Extrusion cooking process for amaranth (*Amaranthus caudatus* L.), *J. Food Sci.*, 65 (6), 1009, 2000.
119. Guzmán-Maldonado, H. and Paredes-López, O., Production of high-protein flour and maltodextrins from amaranth grain, *Proc. Biochem.*, 29 (4), 289, 1994.
120. Escudero, N.L. et al., Comparison of the chemical composition and nutritional value of *Amaranthus cruentus* flour and its protein concentrate, *Plant Foods Hum. Nutr.*, 59 (1), 15, 2004.
121. Rangarajan, A. et al., Iron bioavaillability fom amaranthus species: 2 — Evaluation using hemoglobin repletition in anemic rats, *J. Sci. Food Agric.*, 78 (2), 274, 1998.
122. Prokopowicz, D., Health promoting attributes of amarantus (*Amaranthus cruentus*), *Med. Wet.*, 57 (8), 559, 2001.
123. Gemma, T., Miura, H., and Hayeshi, K., Effects of water depth and temperature on the seedling growth of wild rice, *Zizania palustris*, *Japan. J. Crop Sci.*, 62 (3), 414, 1993.
124. Makela, P., Archibold, O.W., and Peltonen-Sainio, P., Wild rice — a potential new crop for Finland, *Agric. Food Sci. Finl.*, 7 (5–6), 583, 1998.
125. Liu, Q.Q. et al., Formation of panicles and hermaphroditic florets in wild rice, *Int. J. Plant Sci.*, 159 (4), 550, 1998.
126. Hoover, R., Sailaja, Y., and Sosulski, F.W., Characterization of starches from wild and long grain brown rice, *Food Res. Int.*, 29 (2), 99, 1996.

127. Duval, M.R., Wild rice (*Zizanzia palustris*), in *Cereals and Pseudocereals*, Williams, J.T., Ed., Chapman and Hall, London, 1995, 247.

128. Archibold, O.W., Straw residues in wild rice (*Zizania palustris* L.) stands in Northern Saskatchewan, *Can. J. Plant Sci.*, 71 (2), 337, 1991.

129. Ramarathnam, N. et al., The contribution of plant food antioxidant to human health, *Trends Food Sci. Technol.*, 6 (3), 75, 1995.

130. Nriagu, J.O. and Lin, T.S., Trace metals in wild rice sold in the United States, *Sci. Total Environ.*, 172 (2–3), 223, 1995.

131. Kohls, C.L., Percich, J., and Hout, C., Wild rice yield losses associated with growth stage specific fungal brown spot epidemics, *Plant Dis.*, 71 (5), 419, 1987.

132. Gunnell, P.S. and Webster, R., Crown and root rot of cultivated wild rice in California caused by *Phythophthor — Erythroseptica sensu lato*, *Plant Dis.*, 72 (10), 909, 1988.

133. Aiken, S.G., Lee, P., and Stewart, J., *Wild Rice in Canada*, N.C. Press Ltd., Toronto, 1988.

# chapter four

# Millets

Various millet species exist throughout the world. Pearl millet (*Pennisetum glaucum* [L.] R. Br.) and finger millet (*Eleusine coracana* L. Gaertn.) are the most widely spread millets in Africa and in some parts of Asia. Utilized millets also include fonio (*Digitaria* sp.), tef (*Eregrostis tef* [Zucc.] Trotter), barnyard millet (*Echinochloa crus-galli* [L.] P. Beauv.), kodo millet (*Paspalum scrobiculatum* L.), and little millet (*Panicum sumatrense* Roth & Roemer & Schultes). Additionally, proso millet (*Panicum miliaceum* L.) and foxtail millet (*Setaria italica* (L.) P. Beauv.) occur in South America and parts of Asia and Europe The average production of millets used to stand between 28.1 and 30.5 million tons [1], but the data about certified organic production of millet products is not currently available.

The nutritive potential of millets in terms of carbohydrate, protein, and energy values can be compared to that of popular cereals like rice, wheat, or barley. Millets contain the highest percentage of healthy dietary fibers among the cereals [2, 3] and a higher mineral content than rice or wheat. Depending on the species, however, different information is available on millets' proximate composition and processing. Millets contain nonnutritional compounds like phytates, phenols, and tannins, which can contribute to antioxidant activity important to health, aging, and metabolic diseases [4]. Millets' phenolic content and associated antioxidant properties vary among species; milling fractionation and food processing (roasting, boiling, and popping) reduce the activity [5]. Millets are also a gluten-free food, commonly used like porridge (Figure 4.1).

Chemical use in millet production is becoming more and more extensive (in Asia and Africa), though information on this topic in other regions is scarce.

*Figure 4.1* Porridge (kasha) made from proso millet like a side dish.

## 4.1 Proso millet

### 4.1.1 Introduction

Proso millet, often referred to as prose (or common millet, hog millet, broomcorn millet, Russian millet, brown corn, and indian millet), was already being used as food in the early Stone Age. Its primary genetic center is thought to be Middle and East Asia. Proso millet was produced in China and Mongolia in 3000 B.C., and Ancient Greek historians wrote that proso production reached from the Black Sea to the Atlantic. Today, proso is the only millet produced in the U.S., South America, Australia, Japan, and some European countries (including Greece, Hungary). In Asia, proso millet constitutes around 14% of all millet production, and in Africa it is not produced at all. In Slovenia during the years1986 to 1990, an average of 240 hectares were sown and the average grain yield was 1850 kg ha$^{-1}$; in dry 1992, the average yield on 144 sown hectares was 840 kg of grain ha$^{-1}$ [6].

Proso is suitable for food, fodder, and industrial processing. Proso porridge is well known, and in Asian countries, people make bread out of proso flour. Seed and straw can be used as animal fodder, as well. Seed contains approximately 15% of protein, 72% of starch, 0.5% of sugar, 3.9% of fat, 0.5% of cellulose, and 1.2% of minerals. Due to its high starch content, it is used as a raw material in processing industries [2].

## 4.1.2 Botany

The *Panicum* genus includes more than 400 species, but only three are important in production: proso millet (*Panicum miliaceum* L., syn.: *Milium panicum* Mill., *M. esculentum* Moench, etc.), foxtail millet (*Panicum italicum* L., syn.: *Setaria italica* ssp. *maxima* Al., *Panicum germanicum* [Roth] syn.: *Setaria germanica*) and mohar which is mostly used for fodder; therefore, only proso millet and foxtail millet are described.

Land races and related species have thus far been inadequately utilized as genetic resources. They may possess the potential for significant improvements in seed yield as well as future seed quality [7].

Producers often sow land race populations in Slovenia, such as Kornberško proso and Belo strniščno proso cultivars.

### 4.1.2.1 Morphology

Roots are fully developed at the brooming stage and often reach a depth of 1 meter. They receive more intensive moisture than other cereals.

Stems grow between 0.7 and 1.0 m high. The first aboveground node may grow one to five shoots or even as many as 20, with thinner sowing and good provisions. Stem intersection is round and the center is filled with parenchyma. In young plants, the stem is light green in color and covered with hairs.

Proso millet leaves are 65 cm long and wider than other cereals; they are covered with hairs and have no auricles. Inflorescence of the proso millet appears as a grape composite, with spikelets forming the basis of the grape. Inflorescence length reaches 15 to 25 cm, with 10 to 40 side branches attached. Branches' lengths define the form of inflorescence. Proso millet take the form of panicles (Figure 4.2), while foxtail millet form inflorescence like pseudospike. The inflorescence of common millet can be loose, semi-tight, or tight, and tight inflorescence can be flag-like or straight. Flowers are bisexual. In individual spikelets, only one top flower is fertile. Only 20% of the plants can be pollinated; the remaining 80% are self-pollinated. Seeds are round or oval and are surrounded by glumes, and the husked seed is yellow. The weight of 1000 seeds is between 4 and 8 g, and hectoliter weight ranges from 70 to 75 kg hl$^{-1}$. The milling fraction of flour can reach 67 to 84%.

## 4.1.3 Climate conditions

Proso millet, a short-season summer annual, is well adapted to crop production systems in semiarid environments due to excellent water-use efficiency. The growing stage duration is 55 to 115 days, and sometimes longer. Assuming

*Figure 4.2* Inflorescence (panicle) of proso millet.

that the average growing stage of the millet is 80 days, it takes 5 to 20 days for the plant to emerge, 10 to 30 days from emergence to tillering, 16 to 36 days from tillering to brooming, and 30 to 45 days from brooming to maturity.

Growth demands are similar to those for maize, although common millet is more resistant to high summer temperatures (to 38 or even 40°C). Water demands are modest, and extreme transpiration coefficient values are between 140 and 250. Common millet resists drought by the temporary termination of growth and curling of leaves. It will, however, only be able to resist drought until the panicle formation stage; after that point, the demand for water increases.

Proso millet is a short-day plant. Longer nights reduce the period of inflorescence formation, which accelerates ripening for 15 to 20 days with stubble sowing. However, proso needs good light; in cloudy weather, yield is lower. Rows planted in the north-south direction provide up to 10% higher yield, due to better light. Proso grows in all soil textures with a regulated water-air regime, but it prefers sandy-loam soil. It does not tolerate acid soil, and nutrient should be easily accessible.

## 4.1.4   Cultural practice

Due to slow growth after emergence and weed competitiveness, weed-free soil is desired in crop rotation. Combining different production systems with crop rotation, delayed planting, a till and no-till system, and so forth can reduced biomass and seed production of weeds, subsequently eliminating proso millet yield loss. In the case of the Anderson [8] trials, a cultural system reduced biomass and seed production of two pigweed species 85% or more in both tilled and no-till systems. If soil is weed-free, suitable previous crops for millet are legumes and other arables. The millet can also be sown after grass-clover mixtures. Cereals, especially spring cereals, are not the most recommendable precrops. In crop rotation, fox millet and maize should not be planted after each other, due to the joint pest-European corn borer, *Ostrinia nubilalis* (Hubner). Soil is cultivated the same as for main crops or earlier sowing dates and a stubble crop. Proso millet planting date studies [9] show that later planting dates were more heavily infested with European corn borer than were earlier dates. Adjusting the cultivation to destroy weeds and keep soil moist is vital. Rough presowing preparation can be done in autumn; in spring, only presowing machines should be used, if possible. Stubble-crop sowings require plowing after harvesting of the precrop.

Proso demands abundant fertilization due to the weak pumping power of roots. Nutrient uptake in the growing stage is quite uneven because proso can uptake 10% of nutrients until tillering. The highest nutrient uptake appears within 40 to 60 days of the beginning of the tillering stage to the formation of assimilates in grain. The highest demands for N, K and Ca are at the brooming stage, and the highest demand for P takes place at early maturity stages (filling the seeds).

With 100 kg of seed with belonging aboveground vegetation mass the nutrient uptake from soil vary from 3.0 to 3.2 kg of nitrogen, 2.6 to 3.0 kg of phosphorus ($P_2O_5$), and 5.0 to 7.0 kg of potassium ($K_2O$). According to nutrient uptake and nutrients in soil, we can calculate the necessary amount of nitrogen for fertilization. We fertilize with phosphorus, potassium, and part of the nitrogen fertilizers during soil cultivation. Additionally, we fertilize in the tillering stage and also before brooming in cases of acute nutrient deficiency. Stubble millet demands less fertilization.

Seed is sown at the highest 1000-seed weight with good germination and purity. Soil temperature at sowing should be around 10°C at a depth of 2 to 3 cm. It can be sown as a main crop or as stubble crop in early summer. Grain yield, tiller number, and straw weight per plant decreased as plant population and row spacing increased, while seed weight and plant height increased in semi arid environments with increased plant population and row spacing. Interrow spacing can be around 12, 15, or even 20 cm, as with cereals, or it can go as high as 40 to 60 cm. It can be sown into stripes from 10 to 15 cm with interrow stripes up 50 to 60 cm — such spacing is competitive to weed development. In this case, it is better to use interrow stripes for weeding. The producer's goal is to achieve 300 germinated seeds $m^{-2}$. For narrow sowing, 1 to 25 kg of seed $ha^{-1}$ should be used. After stubble sowing, rolling is obligatory.

Agdag et al. [10] report that seed weight increased with row spacing at all row spacings (15, 19, 23, 30, 38, and 46 cm) in the irrigated treatment while on the dry land sites; the increase was noted only at the narrower row spacing. The relatively light weed pressures the effect of weeds in the response of grain yield to row spacing. Narrow rows were superior for grain yield and weed control. In areas of limited moisture, row spacing of 19 cm appeared to be a better choice of row spacing for areas with likelihood of draught.

Besides the necessary rolling, the crushing of crust before emergence is one of the first provision measures. With wide row spacing, weeds can be exterminated in an ecologically friendlier way by hoeing between rows. The first time, we can hoe when rows appear; the second time, 14 days later; and the third time only if needed.

Proso ripens unevenly; consequently, two-phase harvesting is recommended if possible when the middle part of the panicle is still at the stage of waxy maturity.

Main crop yield is a bit higher in comparison with stubble crop, as described at the beginning of the chapter. Intensive production increases the yield from the existing 800–1800 to 5000 kg seed ha$^{-1}$. It can be stored the same as most cereals with at least 14% seed moisture.

### 4.1.5   Utilization

Baby foods made from proso millet like sweet gruel, salty gruel, helwa, burfi, and biscuits based on malting and popping processes were nutritionally similar and organoleptically acceptable [11].

The findings [12] suggest that proso millet protein is considered to be a preventive food for liver injury. Based on mice experiments [13], foxtail millet protein may also serve as a beneficial food component in obesity-related illnesses such as type 2 diabetes and cardiovascular diseases.

## 4.2   Foxtail, foxtail millet

### 4.2.1   Introduction

Foxtail (also called foxtail millet, italian millet, german millet, hungarian millet, and siberian millet (*Setaria italica* [L.] P. Beauv. ssp. *maxima* Alef., syn.: *Panicum italicum* L.) played an important role in Italy, Germany, Hungary, and Siberia, although it originated in East Asia. China is the most important producer of foxtail millet, where it was said to have been produced in 2700 B.C. and still ranks among the five most important crops. Europe probably began producing foxtail millet around the Stone Age. Today, it is the only millet produced in Hungary, Spain, Portugal, and Italy.

### 4.2.2   Biological characteristics

Foxtail millet is said to resemble common millet, though there are many morphological differences between the two. The foxtail millet has a more developed root system: in favorable growing conditions, roots can reach a depth of 1.5 m. Sometimes, adventitious roots can also develop. Plant height ranges between 1 and 1.5 (2) m, and stem length correlates to the length of the growing stage and time of harvesting. Stems are green or colored with

*Figure 4.3* Inflorescences of foxtail millet.

antocyans. Leaves are wide, mostly naked, and are colored the same as stems; there are 6 to 15 of them on one stem.

Inflorescence is tight with short side branches, and the panicle form is similar to an animal tail (Figure 4.3). Inflorescence length varies from 7.5 to 25 cm, with a diameter between1.2 to 5 cm. Spikelets are small, round, or oval and surrounded with hair. Spikelets are made up of of three glumes that embrace two flowers; the largest glume is the outer one, and the smallest inner glume is only an atrophied remnant of the second spikelet. Only the upper flower is fertile. There can be anywhere from 3000 to 5000 flowers in one inflorescence. The plant is a largely self-pollinating species.

Seed color may be yellow, orange, red, brown, or black; the weight of 1000 seeds is approximately 2 g (generally ranges from 1.5 to 4.1 g).

### 4.2.3 Cultivars

Many millet cultivars were selected at University of Nebraska. White Wonder and Golden German cultivars will not usually ripen at unsuitable latitudea and with late sowing (e.g., after June 1 at 43° Northern latitude). Three cultivars with compact cylinders and short inflorescence have been cultivated in Serbia and named after the color of their seed: Novosadski ûuti, Novosadski crveni, and Novosadski crni. Novosadski bar has longer inflorescence. Nutrient content depends on the cultivar; experimental values ranged from 91.7 to 112.0 g kg$^{-1}$ [14]. Seed Resource, Inc. (Tulio) is putting the cultivar German strain R on the market, which is similar to the cultivar Frideriko from Karcag, on the market in Hungary since 1996. New varieties have also been developed, including Yaguane, Carape, and Nandu [15]. However, high-parent heterosis for grain yield with 68% increased yielding over the average yield of the parental cultivars was detected with varietal crosses [16]; therefore, new breeding programs develop new hybrids.

## 4.2.4   Cultural practice

Foxtail millet production is similar to common millet production, but it is sown as a main crop at a temperature range of 8 to 10°C. Interrow spacing should be bigger than with common millet, e.g., 45 to 60 cm. For proper sowing, 6 to 8 kg of seed ha$^{-1}$ are required, or 200 to 300 g of viable seeds m$^{-2}$. Foxtail millet is sown more shallow than common millet; emerging plants grow very slowly to the brooming stage and are poor competition for weeds. Therefore, it is important that the soil is weed-free. Young plants can survive temperatures between –2 and –3°C; foxtail millet is also resistant to drought and high temperatures. The foxtail transpiration coefficient is below 200, a relatively low number.

Millet do not always respond the same way to nitrogen. In general, where precrop provisions are rich with nutrients, foxtail millet production is not recommended. However, it is still necessary to provide nutrients to poor soil and soil with low expected mineralization. The sufficient nutrient amount ranges from 40 to 45 kg of nitrogen ha$^{-1}$, 20 kg P ha$^{-1}$, and 20 kg K ha$^{-1}$.

Genotype × environment interaction is highly significant for grain yield, and bar ripens quite evenly. Harvesting should begin when the inflorescence color is dark.

## 4.2.5   Utilization

Husked seed is a tasty, nutritious porridge that cooks quickly. An edible seed part quantity of 100 g (at 12% moisture) contains 11.2 g of protein, 4.0 g of fat, 3.3 g of ash, 6.7 g of raw fiber, 63.2 g of carbohydrates, 31 mg of Ca, 2.8 mg Fe, 0.59 mg of thiamine, 0.11 mg of riboflavin, and 3.2 mg of niacin [17]. Essential amino acid content is the following (mg g$^{-1}$): isoleucine 475, leucine 1044, lysine 138, methionine 175, phenylalanine 419, threonine 194, tryptophan 61, and valine 431 mg g$^{-1}$ [17]. From a nutritional point of view, it is significant that foxtail millet is gluten-free [2*].

In the case of Ushakumari et al. [18], foxtail millet grains were decorticated in rice-milling machinery, and the decorticated millet was processed to prepare flaked, extrusion-cooked, and roller-dried products, whereas the native grains were subjected to high-temperature, short-time treatment to prepare popped millet. They concluded that the investigated cereal processing technologies could be successfully applied to foxtail millet to prepare ready-to-eat or ready-to-use products, thereby increasing its utilization as a food.

---

* Hulse, J.H., Laing, E.M., and Pearson, O.E., *Sorghum and Millets: Their Composition and Nutritive Value*, Academic Press, New York, 1980, 997 pp.

## 4.3   Pearl millet

### 4.3.1   Introduction

Pearl millet (*Pennisetum glaucum* L.; syn.: *P. typhoideum* [Burm. f.] Rich., *Panicum glaucum* L.) is also called spiked millet, bulrush millet, cat's tail millet, bajra, cumbu, dkhn, sajje, gero, sanio, and souna. Pearl millet is the sixth most important cereal in the world, including pseudocereals. It is the most widely grown millet [1] and investigated of all millets. Pearl millet originated in western Africa, where many wild and cultivated genotypes of millets can be found. More than 2000 years ago, it was also produced in eastern and middle Africa and India.

### 4.3.2   Growth conditions and cultural practice

Pearl millet is well adapted to aridity and high temperatures. However, it also has great potential for use in many regions of Latin America, the Middle East, parts of Eastern Europe, and Central Asia due to hotter and drier climates, especially if the pearl millet is improved for dry conditions.

While pearl millet is well spread in Africa and India, special care should be taken for biological plant protection and appropriate crop rotation, which can help avoid economically harmful pests and diseases. However, in newer potential regions, some of them threats may be avoided.

Pearl millet can be considered one species with many genotypes. The phenology of genotypes has had varying influences on morphological characteristics and yielding ability in contrasting environments. Plants grow between 0.5 and 3 or 4 meters high, depending on genotype. The early cultivars can mature within 60 to 95 days; the latest ones within 120 to 150 days [20]. While early maturation was originally advantageous for stress conditions, cultivars with longer vegetation periods tend to yield more under non-stress conditions. The grain yields of 30 cultivars grown under stress environments ranged between 828 and 1136 kg ha$^1$; under nonstress environments, yields ranged between 3123 and 3942 kg ha$^1$ [19].

Downy mildew, caused by *Sclerospora graminicola*, is a major pathogen of pearl millet (*Pennisetum glaucum*) in Asia and Africa; accordingly, the development of resistant cultivars has been a major goal of both national and international breeding programs, especially for organic production. On the basis of research done by Thakur et al. [20], two lines — 700651 and P 310-17 — could prove to be the most valuable sources of downy mildew resistance. *Pseudomonas fluorescens* may also be used to treat seeds with a pure culture and formulate in talc powder in order to prevent downy mildew disease. Seed treatment turned out to have better results than foliar application alone, but efficacy was significantly higher when seed treatment was followed by a foliar application [21].

*Table 4.1* Proximate Nutritional Composition of Pearl Millet Comparing Different Authors

| Component | Rachie [24] | Hulse et al. [2] | Abdalla et al. [22] |
|---|---|---|---|
| Moisture | 10% | 12% | 9–12% |
| Food energy | 1483 | 1520 | – |
| Major nutrient (%) | – | – | – |
| Protein | 11.8 | 11.8 | 8.5–15.1 |
| Carbohydrates | 70.0 | 67.0 | 58.0–70.0 |
| Fat | 4.8 | 4.8 | 2.7–7.1 |
| Fiber | 1.9 | 2.3 | 2.6–4.0 |
| Ash | 2.3 | 2.2 | 1.6–2.4 |

*Source:* Reprinted from *Cereal Chem.*, 69, Chauhan, G.S., Eskin, N.A.M., and Thachuk, R., Nutrients and antinutrients in quinoa seed, 85, 1992, Copyright 2006, with permission from American Association of Cereal chemists.

*Source:* Reprinted from Hulse, J.H., Laing, E.M., and Pearson, O.E., *Sorghum and Millets: Their Composition and Nutritive Value*. Copyright 2005, with permission from Elsevier.

*Source:* Reprinted from Abdalla, A.A., El Tinay, A.H., Mohamed, B.E., and Abdalla, A.H., *Food Chem.*, 63 (2), 243. Copyright 2005, with permission from Elsevier.

Seeds can be white, yellow, brown, grey, or purple and range in length from 3 to 4 mm, which is shorter than other millets. The 1000-seed weight is between 2.5 and 14 g (8 g on average).

## 4.3.3   Nutritional value

Pearl millet contains 1.6–2.4% ash, 2.6–4.0% crude fiber, 2.7–7.1% oil, 8.5–15.1% crude protein, 58–70% starch (Table 4.1), and 354–796 mg $g^1$ of antinutritional phytic acid. Mineral contents were 10–80, 180–270, and 450–990 mg $g^1$ of Ca, Mg, and P, respectively, and 70–110, 4–13, 53–70, 18–23, 10–18, and 70–180 µg $g^1$ of K, Na, Zn, Mn, Cu, and Fe, respectively. The percentage of phytate to total phosphorus was found to range from 70–89%, with an average of 77%. A linear relationship between phytate and total P existed, with a correlation coefficient of 0.9805 [22*]. According to Hulse et al. [2], approximate values were 42 mg $g^1$ for Ca, 11.0 µg $g^1$ for Fe, 0.38 mg for thiamine, 0.21 mg for riboflavin, and 2.8 mg for niacin. Content of essential amino acids is the following (mg $g^{-1}$): isoleucine 256, leucine 598, lysine 138, methionine 154, cystine 148, phenylalanine 301, tyrosine 203, threonine 241, tryptophan 122, and valine 345 mg $g^{-1}$ [17]. Value (g essential amino acid $16g^{-1}$ N) range amounts are as follows: isoleucine 3.9–4.6, leucine 9.5–12.4, lysine 2.8–3.2, methionine 1.8–2.6, cystine 1.6–1.8, phenylalanine 4.1, tyrosine 3.0, threonine 3.3–4.1, tryptophan 1.4–1.5 and valine 4.9–6.0 g $16 g^{-1}$ N.

---

* Reprinted from *Food Chem.*, 63 (2), Abdalla, A.A. et al., Proximate composition, starch, phytate and mineral contents of 10 pearl millet genotypes, 243, Copyright 2005, with permission from Elsevier.

### 4.3.4 Pearl millet processing and utilization

The traditional foods made from pearl millet are porridge and a beverage product called "fura" (in the Hausa language). Fura is a molded ball of stiff precooked product produced after three to four periods of pounding, alternated with the boiling and molding of the pounded product, and eaten mixed with milk as thick slurry [1]. Malted pearl millet is used for brewing traditional African beer, and traditional flat bread was made from fermented pearl millet flour. The bread baked from pearl millet has unacceptable consistency, due to no gluten fraction in the flour, which is important for cohesive dough. The bread should be made from small particles of pearl millet flour mixed with wheat flour. Other products of the milling process are couscous and porridges blended with other pseudocereals (tef, groundnut, cowpea, amaranths, and so forth).

Pearl and other millets no longer play much of a role in western civilizations. The introduction of pearl millet may raise difficulties, because the flour does not smell good due to one of the C-glycolsyl flavones in the grains [23]; consumers may also avoid the semidark colored products. However, it may be used in the future as an organically produced dietary food. Based on old traditions, new processing possibilities, and new culinary arts, the grains may substitute rice and other grains used as side dishes for meat, porridge, soups, mixed breads, special kinds of beers, and so forth.

## 4.4 Finger millet

### 4.4.1 Introduction

Finger millet (*Eleusine coracana* [L.] Gaertn., syn.: *Cynosurus coracanus* L., *E. stricta* Roxb.) is also known by the following names: African millet, bird's food, coracana, *ragi* in India, *dagusha* in Ethiopia and Eritrea, *wimbi* in Swahili in East Africa, *bulo* in Uganda, and *tamba* and *pwana* in Nigeria.

Finger millet is an underutilized pseudocereal of African or Indian origin [25]. It is an important source of food in Central Africa and India. Production spread to the western parts of southwest India, to the hills under the Himalayas, and to Nepal and along the hills of east China and southern Asia. It is the most important food in Uganda and Zambia. Uganda is considered the primary genetic center, and India began growing millet more than 3000 years ago. According to Vietermeyer [26], the finger millet is a grossly neglected or practically lost crop compared with other cereals, due to its low social esteem. Among the four millets of the world, finger millet ranks fourth and comprised only about 8% in area and 11% in production of all millets in the world during the 1970s [27], with world production at 4.5 million tons [28].

## 4.4.2   Biology

Plants reach from 0.4 to 1.0 m, and panicles are 3 to 13 cm long. They exist in two different genotypes; one is the African highland with seeds closed in the panicles, and the other is an African-Asian genotype with seeds ripening outside the panicles. The 1000 grain-weight is approximately 2.5 g. Grain colors range from light brown to dark brown; certain Nigerian samples of finger millet seeds were roundish and dark reddish-brown, 1 mm in diameter, with a hard interior [29]. The seed coat is bound tightly to the endosperm; its structure is soft and friable.

## 4.4.3   Growth and production characteristics

The finger millet is widely adapted to soil characteristics, and it also tolerates some degree of alkalinity. It is a low-input crop cultivated in arid and poor countries, where the application of complete fallow, manure, and corralling practices are strongly related to a households' endowment of resources, especially with regard to animal production. The plant is well adapted to dry climatic conditions, but its cultivation requires slightly more water than other millets. Annual precipitation range varies from 500 to over 1000 mm. The development of better agricultural practices and improvement of genotypes may be promising, especially when using finger millet as a crop for organic farming. In this case, the target fertilization and optimal soil conditions are possible to maintain with organic fertilizers. Weed management at the first growth stages can be mechanical in interrow spacing, and may sometimes be manual in the rows.

Finger millet achieves the highest yield among millets at 1000 to 1200 kg grain ha$^{-1}$. The grains can be stored for years without pest damage.

## 4.4.4   Nutritional value and utilization

Nutritional value of the seeds is as follows: 100 g of edible part of seed (at 12% moisture) contains 7.7 g of protein, 1.5 g of fat, 2.6 g of ashes, 3.6 g of crude fiber, 72.6 g of carbohydrates, 1406 KJ of energy value, 350 mg of Ca, 3.9 mg of Fe, 0.42 mg of thiamine, 0.19 mg of riboflavin, and 1.1 mg of niacin [34*]. The high fiber content slows the rate of digestion, enabling consumers to work for long hours on a single meal of this millet. The grain is high in protein, fat, and minerals (calcium, iron, and phosphorous), relative to rice, corn, and sorghum [2, 30, 31]. Content of essential amino acids is the following (mg g$^{-1}$): isoleucine 275, leucine 594, lysine 181, methionine 194, cystine 163, phenylalanine 325, tyrosine, threonine 263, tryptophan 191, and valine 413 mg g$^{-1}$ [17]. However, it is also an excellent dietary source of methionine, an essential amino acid [32]. Evidence has long shown that patients with

---

* Reprinted from *Sorghum and Millets: Their Composition and Nutritive Value*, Hulse, J.H., Laing, E.M., and Pearson, O.E., 997 pp, Copyright 2005, with permission from Elsevier.

diabetes tolerate finger millet better than rice, and that their blood sugar levels are lower [27].

The finger millet was traditionally used as porridge made from boiled groats; foods made from the flour are mainly mixed with cereals, and it is also used for malting in beer substitutes.

Food processing from finger millet was investigated and, in few cases, results support applicability. However, the heat-moisture treatments modify the finger millet starch pasting properties of the native starch from type B, which is a characteristic of normal cereal starches, to type C, which is a characteristic of cross-linked or legume starches, while the annealing starch retains its type C configuration. In both cases, the damage of starchy granules was lower than 1% [33**]. Fermented and nonfermented thin porridges can also be made from millets, as from buckwheat. For finger millet, this reportedly leads to an over 6% reduction in starch content during a 48-hour fermentation period [34]. Of significance to the rheology of cooked paste is the amylolytic degradation of starch in the fermenting slurry. Nelson et al. [35] described that the effects of flour solid concentration, cooking time, and temperature on yield stress and apparent viscosity of fermented finger millet pastes displayed biphasic temperature dependence in samples boiled for 60 minutes, with the critical temperature at around 40°C.

The germination and sprouting of finger millet seeds have significant effects on compound changes. The sprouted finger millet at 30°C in well-ventilated vessels decreases in antinutritional factors, with tannins and phytates [36] decreasing to undetectable levels. Trypsin inhibitor activity decreased threefold. By the end of the optimal 48 hours of germination, there were high decreases in viscosity and starch content accompanied by large increases in sugar content. Trypsin inhibitor activity significantly decreased, and significant changes also occurred in *in vitro* protein digestibility [37, 38].

## 4.5 White fonio

### 4.5.1 Introduction

Fonio is the tastiest and one of the most underutilized pseudocereals. Their genotypes for further development have not yet been improved enough, nor utilization of the crop studied enough. The best description of fonio was done by Haq and Dania Ogbe [39] and Babutande and Maqnyasa [1].

### 4.5.2 Biology

Fonio has been cultivated across dry savannas for thousands of years and once represented a major food source in western Africa; its production was documented in west Sudan from 6500 to 2500 B.C. The *Digitaria* genus

* Reprinted from *Food Hydrocoll.*, 19 (6), Adebowale, K.O., Afolabi, T.A., and Olu-Owolabi, B.I., Hydrothermal treatments of Finger millet (*Eleusine coracan*) starch, 974, Copyright 2005, with permission from Elsevier.

consists of 230 to 325 species, but only white-grained fonio *Digitaria exilis* Stapf with geographical and morphological varieties var. *gracilis*, var. *stricta*, var. *rustica*, var. *rubra*, and var. *densa* and black fonio *Digitaria iburua* grown in parts of Nigeria, Togo, and Benin are important for production purposes. Within these many varieties, there are 15 agronomically recommendable cultivars with significant characteristics of these ecotypes.

Fonio grows about 45 to 50 cm in height and has finger-like panicle that reach up to 15 cm in length. More primitive cultivars contain 50 to 100 spikelets per 10 cm of panicle, whereas more refined ones contain 120 to 140 spikelets per 10 cm of panicle length. The grain may be white, yellow, or purple. The seeds are very small, with 2500 to 3000 grains weighing about 1 g.

Early genotypes have very short vegetation periods and can mature 6 to 8 weeks after sowing, especially when sown after rainy periods in arid regions. Fonio can produce grain in rainfall totaling under 300 mm and more than 3000 mm. Fonio tolerates very poor and degraded soils, but for successful organic production, the guidelines of good agricultural practice should be followed, especially with regard to the care of organic substances in the soil, their microbiological activity, and nutritional balance.

### 4.5.3   *Cultivation and utilization*

Fonio is either sown as a sole crop or intercropped with pearl millet, sorghum, or millet. In the case of intercropped crop, the maturity stages of both crops should take place during the same period.

The food made from fonio is very tasty — it is considered to be one of the world's best tasting and nutritious pseudocereals. It is used for special occasions like ancestor worship and as the price paid for a bride [1]. Flour from its small seeds is very useful and can replace poppy seeds or malt in beer production; can be served as a side dish with meat or spinach; or can be added to soups. In Nigeria, it is consumed three times a day in various forms produced from many different cultivars.

Yield from this C4 plant ranges from 170 to 2000 kg grain ha$^{-1}$, but in West African countries with an average of 200 to 500 kg grain ha$^{-1}$ [40], the normal range should be between 600 and 1400 kg ha$^{-1}$.

According to Maleshi [41], 100 g of seeds contains 7.2 to 8.7 g of crude proteins, 1.1 to 1.8 g of fat, 1.3 to 1.6 g of fiber, and 74.4 to 81.0 g of carbohydrates. Sources [24, 41] have reported fonio to have the following range of mineral content: 30–44 mg of Ca 100 g$^{-1}$, 16 mg of K 100 g$^{-1}$, 40–44 g of Mg 100 g$^{-1}$, 8.5 mg of Fe 100 g$^{-1}$, and 170–177 mg of P 100 g$^{-1}$ [41, 42]. An average protein content found in the black fonio varies from 6 to 11.8% [1]. Content of essential amino acids is as follows (g g$^{-1}$ N): isoleucine 0.25, leucine 0.61, lysine 0.16, methionine 0.35, cystine 0.13 to 0.18, tyrosine 0.023, threonine 0.21 to 0.27, tryptophan 0.09, and valine 0.34 mg g$^{-1}$ [40]. Methionine levels are higher than those found in other millets and sorghum, and twice as high as what is found in egg protein [1]. Values for thiamin, niacin, and riboflavin are 47, 190, and 10 mg 100 g$^{-1}$, respectively [24].

No notes currently exist about the development of fonio processing and parallel research. However, the traditional techniques are well developed and enable processing of organic products, especially health and snack foods. Standard postharvest technology after threshing the grains include sieving, winnowing, washing, and drying to the 13% moisture in grains. The grains are eaten as a popped product; they are mostly ground or crushed into flour to make couscous or porridge. From a mixture of other millets, sorghum, and wheat flour, a composite bread can be made. The flour can be used as a substitute for semolina when making wheat noodles, pastas, and spaghetti. The grain is also malted in small breweries for processing local beers in Togo and Nigeria [1].

## 4.6 Barnyard millet

Barnyard represents the two follows species:

1. Barnyard grass, (wild) millet, pricky grass (*Echlinochloa* [*Panicum*] *cruss-gali* [L.] P. Beauv.)
2. Japanese barnyard millet, sham millet, jungle rice (*E.* [*P.*] *colona* [L.] Link var. *frumentacea* Blatter et Mc Cann)

It is one of the fastest growing cereals and can be produced in 6 weeks, as fast or faster than fonio millet. Barnyard is one of 8 millet species and among the least produced. In India, Japan, and China, it is grown as a rice substitute when husked rice is lacking. It may also be produced as fodder in some places up to eight times in a year.

Plants reach a height of 0.5 to 1 m. The grain yield is modest but with further improvements more productive genotypes could be created. In organic farming, fast-growing barnyard can cover field surfaces even in dry periods of the year. It is also a good link in crop rotation.

Edible seed parts totaling 100 g (at 12% moisture) contain 11.0 g of protein, 3.9 g of fat, 4.5 g of ash, 13.6 g of raw fiber, 55.0 g of carbohydrates, 1265 KJ of energy value, 22 mg of Ca, 18.6 mg of Fe, 0.33 mg of thiamine, 0.10 mg of riboflavin, and 4.2 mg of niacin [2]. Essential amino acid content is as follows (mg $g^{-1}$): isoleucine 288, leucine 725, lysine 106, methionine 133, cystine 175, phenylalanine 362, tyrosine 150, threonine 231, tryptophan 63, and valine 388 mg $g^{-1}$ [17].

According to its nutritional value, it is a promising crop for organic agriculture and special products, just like other millets.

## 4.7 Koda millet

Koda millet, also known as dith millet and koda kodra (*Paspalmus scrobiculatum* Lam., syn. *P. kodra* Wild., *Panicum dissectum* L.), is tuff grass spread in India that reaches a height of up to 90 cm. It is divided into four categories according to the inflorescence form. Due to the infestation of seeds with fungi, some categories may be poisonous for humans. Grain is hard and hornlike with very resistant epidermis. The color of the grain varies from light red to dark grey.

Edible seed parts totaling 100 g (at 12% moisture) contain 9.8 g of protein, 3.6 g of fat, 3.3 g of ash, 5.2 g of raw fiber, 66.6 g of carbohydrates, 1477 KJ of energy value, 35 mg of Ca, 1.7 mg of Fe, 0.15 mg of thiamine, 0.09 mg of riboflavin, and 2.0 mg of niacin [2]. Essential amino acid content is as follows (mg g$^{-1}$): isoleucine 188, leucine 419, lysine 188, methionine 94, phenylalanine 375, tyrosine 213, threonine 194, tryptophan 38, and valine 238 mg g$^{-1}$ [17].

## *4.8. Little millet*

Little millet, also called small millet (*Panicum sumatrense* Roth & Roemer & Schultes, syn.: *Panicum miliare* Lam.), is produced in India up to 2100 m above sea level. The production effect of little millet is low but it still catches the attention of plant breeders. Plants grow between 0.3 and 0.9 m high, and the inflorescence length ranges from 14 to 40 cm. Seed is much smaller than that of common millet seed.

Edible seed parts totaling 100 g (at 12% moisture) contain 9.7 g of protein, 5.2 g of fat, 5.4 g of ash, 7.6 g of raw fiber, 60.9 g of carbohydrates, 1377 KJ of energy value, 17 mg of Ca, 9.3 mg of Fe, 0.30 mg of thiamine, 0.09 mg of riboflavin, and 3.2 mg of niacin [53]. Essential amino acid content is as follows (mg g$^{-1}$): isoleucine 416, leucine 679, lysine 114, methionine 142, phenylalanine 297, threonine 212, tryptophan 35, and valine 379 mg g$^{-1}$ [17].

## *References*

1. Babatunde, O. and Manyasa, E., Millets, in *Pseudocereals and Less Common Cereals: Grain properties and Utilization Potential*, Belton, P. and Taylor, J., Eds., Springer-Verlag, Berlin, 2002, 177.
2. Hulse, J.H., Laing, E.M., and Pearson, O.E., *Sorghum and Millets: Their Composition and Nutritive Value*, Academic Press, New York, 1980.
3. Malleshi, N.G and Hadimani, N.A., Nutritional and technological characteristics of small millets and preparation of value-added products from them, in *Advances in Small Millets*, Riley, K.W. et al., Eds., Oxford & IBH Publishing Co. Pvt. Ltd., New Delhi, 1993, 271.
4. Bravo, L., Polyphenols: Chemistry, dietary sources, metabolism and nutritional significance, *Nutr. Rev.*, 56, 317, 1998.
5. Prashant Hegde, S. and Chandra, T.S., ESR spectroscopic study reveals higher free radical quenching potential in kodo millet (*Paspalum scrobiculatum*) compared to other millets, *Food Chem.*, 92 (1), 177, 2005.
6. Bavec, F., Millets, in *Nekatere zapostavljene in/ali nove poljščine (Some of disregarded and/or new field crops)*, Univerza v Mariboru, Fakulteta za kmetijstvo, Maribor, 2000, 186 pp.
7. Zeller, F.J., Utilization, genetics and breeding of small-seeded millets: 3. Proso (*Panicum miliaceum* L.), *J. Appl. Bot.*, 74 (5–6), 182, 2000.

8. Anderson, R.L., A cultural system approach can eliminate herbicide need in semiarid proso millet (*Panicum miliaceum*), *Weed Technol.*, 14 (3), 602, 2000.

9. Anderson, P.L. et al., Millet preference, effects of planting date on infestation, and adult and larval use of proso millet by *Ostrinia nubilalis* (*Lepidoptera: Crambidae*), *J. Econ. Entomol.*, 96 (2), 361, 2003.

10. Agdag, M. et al., Row spacing affects grain yield and other agronomic characters of proso millet, *Commun. Soil Sci. Plant Anal.*, 32 (13–14), 2021, 2001.

11. Srivastava, S., Thathola, A., and Batra, A., Development and nutritional evaluation of proso millet-based convenience mix for infants and children, *J. Food Sci. Technol.*, 38 (5), 480, 2001.

12. Nishizawa, N. et al., Effects of dietary protein of proso millet on liver injury induced by D-galactosamine in rats, *Biosci. Biotechnol. Biochem.*, 66 (1), 92, 2002.

13. Choi, Y.Y. et al., Effects of dietary protein of Korean foxtail millet on plasma adiponectin, HDL-cholesterol, and insulin levels in genetically type 2 diabetic mice, *Biosci. Biotechnol. Biochem.*, 69 (1), 31, 2005.

14. Kumar, K.K. and Parameswaran, K.P., Characterisation of storage protein form from selected varietes of foxtail millet (*Setaria italica* [L.] Beauv.), *J. Sci. Food Agric.*, 77 (4), 535, 1997.

15. Pallares, I.N. et al., Identification of cultivars of foxtail millet (*Setaria italica* [L.] Beauv.) from morphological and biochemical traits of their seed, *Seed Sci. Technol.* 32 (1), 187, 2004.

16. Siles, M.M. et al., Heterosis for grain yield and other agronomic traits in foxtail millet, *Crop Sci.*, 44 (6), 1960, 2004.

17. Indira, R. and Naik, M. S., Nutrient composition and protein quality of some minor millets, *Indian J. Agric. Sci.*, 41, 795, 1971.

18. Ushakumari, S.R., Latha, S., and Malleshi, N.G., The functional properties of popped, flaked, extruded and roller-dried foxtail millet (*Setaria italica*), *Int. J. Food Sci. Technol.*, 39 (9), 907, 2004.

19. Yadav, O.P. and Bhatnaga, S.K., Evaluation of indices for identification of pearl millet cultivars adapted to stress and non-stress conditions, *Field Crops Res.*, 70 (3), 201, 2001.

20. Thakur, R.P. et al., Host resistance stability to downy mildew in pearl millet and pathogenic variability in *Sclerospora graminicola*, *Crop Prot.*, 23 (10), 901, 2004.

21. Umesha, S. et al., Biocontrol of downy mildew disease of pearl millet using *Pseudomonas fluorescens*, *Crop Prot.*, 17 (5), 387, 1998.

22. Abdalla, A.A. et al., Proximate composition, starch, phytate and mineral contents of 10 pearl millet genotypes, *Food Chem.*, 63 (2), 243, 1998.

23. Hoseney, R.C., Faubian, J.M., and Reddy, V.P., Organoleptic implications of pearl millet, in *Utilization of Sorghum and Millets*, Gomez, M.I. et al., Eds., ICRISAT, Patancheru, India, 1992, 27.

24. Rachie, K.O., The millets: Importance, utilization and autlook, ICRISAT, Hyderabat, 1975,1.

25. Watt, J.M. and Breyer-Brandwijk, M.G., *The Medicinal and Poisonous Plants of Southern and Eastern Africa*, 2nd Ed., Livingstone, Ltd., Edinburgh, 1962.

26. Vietmeyer, N.D., *National Academy of Science: Lost Crops of Africa, Grains*, Vol. 1, Vietmeyer, N.D., Ed., National Academy Press, Washington, DC, 1996, 39.

27. Rachie, K.O. and Peters, L.V., *Pseudocereals and Less Common Cereals: Grain Properties and Utilization Potential*, Belton, P., and Taylor, J., Eds., Springer-Verlag: Berlin, 2002.

28. Rao, M.V., The millets: Their inportance, present status and autlook, in *Small Millets in Global Agriculture*, Proceedings of the first International small millets workshop, Bangalore, India, Oct. 29–Nov. 2, 1986, Seetharam, A., Riley, K.V., and Harinayarana, G., Eds., ICRISAT, Bangalore, India, 1989, 9.

29. Bond, B. et al., Fatty acid, amino acid and trace mineral analysis of three complementary foods from Jos, Nigeria, *J. Food Compos. Anal.*, 18 (7), 675, 2005.

30. Reed, C.F., *Information Summaries on 1000 Economic Plants*, Typescripts submitted to the USDA, 1976.

31. Subbarao, M.V.S.S.T. and Muralikrishna, G., Non-starchy polysaccharides and bound phenolic acids from native and malted finger millet (*Eleusine coracana* Indaf-15), *Food Chem.*, 72, 187, 2001.

32. Duke, J.A., Ecosystematic data on economic plants, *Q. J. Crude Drug Res.*, 17, 91, 1979.

33. Adebowale, K.O., Afolabi, T.A., and Olu-Owolabi, B.I., Hydrothermal treatments of Finger millet (*Eleusine coracan*) starch, *Food Hydrocoll.*, 19 (6), 974, 2005.

34. Antony, U., Sripriya, G., and Chandra, T.S., Effect of fermentation on primary nutrients in finger millet (*Eleucine coracana*), *J. Agric. Food Chem.*, 44, 2616, 1996.

35. Nelson, K.O.O. and Eyal, S., Rheological properties of fermented finger millet (*Eleucine coracana*) thin porridge, *Carbohydr. Polym.*, 57 (2), 135, 2004.

36. Lorenz, K., Tannins and phytate content in proso millets (*Panicum miliaceum*), *Cereal Chem.*, 60, 424, 1983.

37. Kumar, H. and Chauhan, B.M., Effects of phytic acid on protein digestibility (*in vitro*) and HCl extractability of minerals in pearl millet sprouts, *Cereal Chem.*, 70, 504, 1993.

38. Mbithi-Mwikya, S. et al., Nutrient and antinutrient changes in finger millet (*Eleusine coracan*) during sprouting, *Lebens.- Wiss.+Technol.*, 33 (1), 9, 2000.

39. Haq, N. and Dania Ogbe, F., Fonio (*Digitaria exilis* and *Digitaria iburua*), in *Cereals and Pseudocereals*, Williams, J.T., Ed., Chapman and Hall, New York, 1995, 175.

40. Mybye, N. and Nwasike, C., Fonio millet (*Digitaria exilis* Stapf) in West Africa, in *Advances in Small Millets*, Gupta, K.W., Seetharam, A., and Mushonga, J.N., Eds., Oxford and IBH Publishing, New Delhi, 1993, 85.

41. Malleshi, N.G., Processing of small millets for food and industrial uses, in *Small Millets in Global Agriculture*, Seetharum, A., Riley, K.W., and Hannardyana, G., Eds., Oxford and IBHS Pubs. Co. Put. Ltd., New Delhi, 1989, 325.

42. Porteres, R., The minor cereals of the genus *Digitaria* in Africa and Europe, *J. Agric. Crop Bot. Appl.*, 2, 349, 1955.

# chapter five

# Alternative oil plants

## 5.1 Oil (seed) pumpkins

### 5.1.1 Introduction

Oil (seed) pumpkin (squash, marrow, oil-bearing gourd) is also known as oil pumpkin. The term was established by Austrian and Slovenian farmers from the Styria region where unique cultivars of field pumpkins (*Cucurbita pepo* L. group Pepo) have been grown for over 100 years. The name was given because of the high percentage of oil in the seeds. Hull-less or naked seeds are edible and easily crushed to extract edible oil, mainly used for salad dressing. Seeds and oil can be used in pharmacology, cosmetics, and alternative medicine, especially when organically produced. Oil content in the seeds is 40 to 60%; the traditionally processed oil is dark green and contains free fatty acids. The content of vitamin E (gamma-tocopherols), additionally, is very high [1]. The seeds and oil, organically produced in some places, are an important product in gastronomy.

### 5.1.2 Botany and growth stages

Oil pumpkins belong to the family of *Cucurbitaceae*, which contains 90 orders and 750 species. The species of squash pumpkin (*Cucurbita pepo* L.) is one of 5 cultivated and approximately 10 wild species of the genus *Cucurbita* L. of family "Cucurbits" *Cucurbitaceae*. Archeological evidence places wild populations of *C. pepo* in Mexico and the eastern U.S., with the oldest evidence dating as far back as 30,000 years. Countries such as Turkey, Little Asia, India, Burma, Japan, and China are considered "secondary centers" of *Cucurbitaceae* divergence [1]. Today, *C. pepo* is grown throughout the world, and includes pumpkins, ornamental gourds, and vegetable marrows.

    *C. pepo* was the first squash introduced to Europe. During European domestication, people selected oil pumpkins containing seeds rich in oil. Two main groups of oil pumpkin varieties currently exist in the main production region of Styria: *C. pepo* L. ssp. *pepo* var. *styriaca* Greb (syn. *C. pepo* convar. *citrullina* [L.] Greb. var. *styriaca* Greb.) and *C. pepo* L. ssp. *pepo* var. *oleifera* Pietsch (syn. *C. pepo* L. var. *oleifera* Pietsch) [2*].

The seeds are naked seeded or hull-less: they have a thin membranous testa (seed coat) rather than the common lignified testa of pumpkin seeds.

Pumpkins develop medium deep taproot, and their roots have weak suction abilities. The plant is of a vine type in most cases; however, bush type plants have been grown at University of Maribor (Ivani) and in Austria. For now, the only cultivars established in production are middle stemmed, such as Gleisdorfer Ölkürbis, Wies 371, Slovenska golica, and Olinka.

The trailing stems of pumpkins are called vines and are square and juicy in diameter. Pumpkins are divided in short and long types according to vine length. Oil pumpkins are mostly short vine types; from the stems grow adventitious roots, tendrils, and branches. The leaves are large with hollow petioles and are covered with prickly outgrowths, and their surface is palmately lobed. Differentiation between pumpkin species is possible because of the varying shapes of leaf sinuses.

Pumpkins are generally monoecious, with separate male and female flowers. Genotypes with unisexual flowers or hermaphroditic varieties also exist. Male flowers are long-stalked, and female flowers have inferior ovary. Most cultivars are able to self-pollinate; pumpkins, however, are insect pollinated. The flowering stage of individual plants is very long and usually takes more than two months.

The fruit is divided into three parts and covered by a harder epidermis with edible layer below. The inner part contains fruit tissue with seeds (Figure 5.2); the seeds of oil pumpkins are naked seeded. Seeds represent approximately 3% of fruit mass.

The phenological stages of oil pumpkins are as follows:

- Germination (dry seed, radicle emerged from the seed, cotyledons visible), leaf development from first true leaf (Figure 5.3) to n-true leaf (Figure 5.4)
- Formation of side shoots (first to n-primary side shoots visible, first to n-secondary side shoots visible)
- Inflorescence emergence (first to n-flower initial with elongation ovary visible on main stem, first flower initial with elongation ovary visible on secondary stem, and so forth)
- Development of fruits (first fruit on main stem has reached typical size and form first, second, and third fruits on main stem have reached typical size and form, first fruit on secondary side shoot has reached typical size and form, and so forth)
- Stages of maturity of fruits and seeds (10 to 80% of fruits show typical fully ripe color; fully ripe fruits show typical fully ripe color; full maturity of harvested seeds vs. shoots, and leaves dead)

* Reprinted from *Cucurbita pepo* (*Cucurbitaceae*) — history, seed coat types, thin coated seeds and their genetics, Teppner, H., *Phyton — Annales Rei botanicae*, 40 (1), 1, with permission from Berger & Söhne.

*Figure 5.1* Oil (seed) pumpkins before harvest.

*Figure 5.2* Oil (seed) pumpkin fruit tissue with seeds.

### 5.1.3   Production

Pumpkins are grown as oil plants (Figure 5.1) only in a small area of central Europe (the Styria region in Austria and Slovenia, the eastern part of Slovenia, the western part of Hungary, the northern part of Croatia, and the northern part of Serbia). Average yielding ranges between 300 and 1200 kg

*Figure 5.3* Oil (seed) pumpkin at cotyledons and first true leaf stage.

*Figure 5.4* Oil (seed) pumpkin at 10–12 true leaves stage.

dry seeds ha$^{-1}$, depending on climatic circumstances, production systems, and the appearance of plant diseases. China is the world's main exporter of edible pumpkins seeds. However, the production of pumpkin seeds is currently on the increase; research on organic production techniques should therefore be encouraged.

## 5.1.4   Soil and climate

Light sandy-loam soil is most suitable soil for oil pumpkins, but they can also be successfully grown in heavier soil. The soil must be rich in humus, airy, and well fertilized with organic substances, so that many nutrients are available to the plants. Pumpkin demands are similar to those of other arable crops. Soil pH should be no lower than 6; in the case of 6 or lower pH, soil calcification should be performed at least a month before sowing. Ploughing depth is adjusted to the depth of the common tillage layerbut should go no deeper than 25 cm. Compact soils and soils that do not have enough drainage are not suitable for pumpkin production. The lack of air at the roots can be avoided in humid conditions by making cards that follow the use of ripe manure in rows, in case fertilization does not occur in autumn. Dry soil is harrowed in spring, as soon as the conditions allow it to preserve as much moisture as possible. Shallow spring plowing and harrowing are executed only if soil is too compact. If suitable air and moisture conditions and the beneficial effects of winter moisture on harrow allow the use of fertilizers in the spring before sowing, sowing can be prepared with only presowing tools. For manual sowing, sowing holes are marked. Fertilization occurs in autumn with 25,000 to 45,000 kg ha$^{-1}$ of stable manure, or the highest amount pre-scribed by regional legislation. In Europe, the annual amount of nitrogen with organic fertilizers should not exceed 170 kg N ha$^{-1}$.

Pumpkin growth and development are aided by a warm climate (approximately 2500°C at the growing stage), high levels of light (exposure to sun), and sufficient soil moisture. Pumpkins are very sensitive to low temperatures: they germinate at 8°C, and growth stops by 6 to 7°C. Iftem-peratures between 2 and 4°C last for longer than 3 days, yield can be reduced by more than half. At a temperature of −1°C, the plants are ruined. Despite their high transpiration coefficient, pumpkins are quite resistant to drought. Nevertheless, yield is significantly smaller in this case.

## 5.1.5   Cultural practice

Oil pumpkins have no special presown crop demands and are a suitable presown crop to almost all other crops. They share diseases only with cucum-bers, so therepeated production of pumpkins or cucumbers in the same field is not recommended for at least 4 to 5 years. Pumpkin production is most successful following legumes, green fodder, green manure, and clover. A high level of organic fertilizers or resistant humus and catch crops that do not winter are good growth conditions for pumpkins.

Pumpkins display the usual pumpkin diseases; except for *Erysiphe polyphaga* and *Sphaerotheca fuliginea*, diseases appear only occasionally and to a limited extent. *Pseudomonas lachrymans* appear occasionally during rainy seasons in fields that are in proximity to larger nurseries. Spraying with copper preparations is not efficient, however. The extent of infection at emergence can be reduced by presowing soil preparation for a light soil structure and abundant organic fertilizing; these precautions will also help reduce any  shriveling or rottenness of the stem caused by *Fusarium and Sclerotinia* fungi. Compost of remains of tomato, pepper, eggplant, cucumber, and bean stems are unsuitable for pumpkin fertilization.

Oil pumpkins are extremely susceptible to the cucumber mosaic virus. It is highly transmittable from aphids to the plant, and partially through the seed. Leaves show blister-like patterns and partially deformed fruits display bump-like bulges. The worst effect of the infection is reduced flowering and flower abortion. Decoration plants, tomato, and pepper are often infected. If the aphids are able to suck sap from plants, pumpkin yield may be highly infected.

Soil suitable for pumpkins is rich in humus and nutrients. When fertilizing, nutrients that have already been added with stable manure must be taken into consideration, and the necessary amount of nutrients required according to uptake and soil nutrient content must be calculated accordingly. We have established the following target values for nitrogen fertilization, based on Slovenian growing conditions, experiments conducted in sandy and sandy-loam soil, and chemical analyses of soil samples in the middle of May: kg N ha$^{-1}$ = 200 kg N ha$^{-1}$ – kg N min ha$^{-1}$ (0 to 0.4 m depth) [3]. Target value was confirmed as suitable with later field microexperiments conducted by students despite high result dispersion.

Plants' uptake was an even 420 kg $K_2O$ ha$^{-1}$, 48 kg $P_2O_5$ ha$^{-1}$, four times the amount of CaO compared to sugar beet (282 kg ha$^{-1}$, only 5% in yield), MgO 43 kg ha$^{-1}$ [4]. Therefore, high levels of nutrients should be available during the vegetation period.

Sowing can be implemented on nongerminated seeds, germinated seeds, and transplants. Seedlings without developed roots are not transplanted, as their transplantation is mostly unsuccessful. Soil temperature should be at least 12 to 15°C at the early growth stage. Pumpkins are sown 3 to 5 cm deep, but the sowing depth should be deeper in sandy soil. Optimal yield may be achieved (with experiments) in cases of crops thinned from 1.0 to 1.5 plants m$^{-2}$. In Slovenian conditions, the traditional recommendation with manual seeding was to plant in the shape of a 1 × 1 m square. Due to the possibility of seeding pumpkins with a pneumatic corn seeder, and based on the later interrow hoeing, we recommend the interrow spacing to be 1.4 m or 2.1 m, with the longest possible spacing between rows. With manual seeding, 3 to 4 kg of seeds ha$^{-1}$ are needed, and  between 6 to 7 kg of seeds ha$^{-1}$ are required for mechanical seeding.

It is important to use a healthy seeding material to avoid any virus diseases that may have decimated the pumpkin yield in previous years. Empty spaces originating from manual square seeding are filled in with germinated pumpkin seeds a week after emergence. Two plant seeds can be placed together and thinned to one plant after emergence — not by pulling but by cutting, in order to avoid damaging adjoining plants. The suggested final plant population for optimal yielding varies between 1 and 1.5 plants m$^{-2}$.

For successful crop provisions, efficient weed control should be provided through manual hoeing and with the use of a mechanical hoe in the spaces between rows. This also helps with moisture preservation, especially if soil is unstructured and crusty. Mechanical hoeing is used as long as the length of stems allows it; at this stage, pumpkins can be hoed two to three times.

Higher yield and more reliable production of cucurbits are often obtained through the use of transplants, not by direct seeding. Plants developed from direct seeding grew and developed more slowly in comparison to plants developed from transplants.

At the University of Maribor transplants were produced in plug trays in a heated greenhouse and seedlings were hardened before transplanting. Plug trays ($28 \times 50$ cm) had 72 holes, each $3.7 \times 4$ cm in size, and 5 cm deep. The volume was approximately 5.5 cm$^3$, because plants should be strong and hardened without the additional shock of stem elongation due to competitiveness resulting from a higher number of plants. Plug trays were filled with the organic substrate with one seed in each hole. Seeds emerged in well-lit areaa at day temperatures ranging from 20 to 25°C and night temperatures of 15°C. Five days after emergence, plants were grown at 20°C. Seedlings need to be grown in optimal conditions in covered spaces and should not endure temperature drops below 10°C; they should also not elongate as much as cucumbers. After 14 days at the first true leaf stage, plants were transferred outdoors and covered with plastic overnight. Seedlings were transplanted after a month to the field location when they had 2 to 3 true leaves, stood 15 cm tall, and displayed visible first flower buds. Seedlings were well developed and roots survived transplantation without damage [5].

Plants developed from direct seeding grew and developed more slowly than the plants developed from transplants. Many morphological differences between the treatments appeared at the beginning of summer. Years and treatments had significant influence on fruit yield, number of fruits, diameter of fruits, and seed yield (Table 5.1). In all years, the transplants had the greatest fruit yield, the highest number of harvested fruit, the largest fruit,

*Table 5.1* Effects of Direct Seeding (Nongerminated and Germinated) and Transplanting on Seed Yield and Yield Characteristics of Oil Pumpkin

| Treatment | Fruit Weight (t ha$^{-1}$) | No. Harvested (unripe/decayed) Fruits ha$^{-1}$ | Fruit Diameter (cm) | Seed Yield (t ha$^{-1}$) (Average) |
|---|---|---|---|---|
| Nongerminated seed | 9.6–44.0 | 4350–11100/0–2500 | 17.1–21.6 | 0.44–1.50 (1.03) |
| Germinated seed | 11.2–51.0 | 3750–12750/150–5250 | 19.0–21.5 | 0.80–1.77 (1.27) |
| Transplants | 30.6–72.6 | 8150–17150/0–5500 | 21.9–22.3 | 1.30–2.36 (1.68) |

*Source:* Reprinted from *Die Bodenkultur,* 53 (1), Bavec, F. et al., Seedlings of oil pumpkins as an alternative to seed sowing: yield and production costs, 39, Copyright 2006, with permission from Die Bodenkultur.

*Figure 5.5* Combine for harvest oil (seed) pumpkins.

and the highest seed yield. There were slight differences between the two seeding treatments, and many were not significant. The average seed yields for three years were 1.03 t ha$^{-1}$ for the direct seeding of dry seed, 1.27 t ha$^{-1}$ for the direct seeding of germinated seed, and 1.68 t ha$^{-1}$ for transplants [5].

## 5.1.6   Harvesting, handling, and storage

To achieve the highest possible share of oil, fruit should be completely mature. Mature fruits are yellow with exception of the shady side, where they have yellow and green stripes. Leaves are yellow at this stage. With mature fruits, cucurbits are firm and the innermost fruit tissue can be easily removed from the seeds, either by hand or machine (Figure 5.5). Especially when mechanically cleaning and drying a larger quantity of seeds, it is necessary to wash them with a lot of water under low pressure; otherwise, the gentle skin and grin part of the rind can be damaged and, consequently, the quality of yield reduced. Seed is dried at 40 to 60°C until it reaches the final moisture content of 8 to 10%. The usual yield of cucurbits is between 600 to 1000 kg ha$^{-1}$, though very successful producers can achieve a yield of 1300 kg of naked seeds of pumpkins ha$^{-1}$ with 9% of moisture [1].

## 5.1.7   Nutritional value

Pumpkin oil is dark green and contains free fatty acids. Vitamin E content, especially gamma-tocopherol, is very high [6–8]. The oil content of the

*Figure 5.6* Pumpkin oil and seed packages.

pumpkin seeds varies from 40 to 50 (54)%. The dominant composition of fatty acids is made up of linoleic acid (C18:2, 35.6–60.8), oleic acid (C18:1, 21.0–46.9%), palmitic acid (C16:0, 9.5–14.5%), stearic acid (C18:0, 3.1–7.4%), and others; the content is below 0.5% fatty acids [9]. The Iodine number of oil is 120 to 130 [10].

### 5.1.8    Gastronomy and health value

The seed oil is used for salad dressings but also has uses in pharmacology and alternative medicine [11], especially when produced organically.

The dry seeds are eaten fresh (Figure 5.6) or roasted. The roasted seeds can be consumed in combination with salt or coated with caramel, chocolate, cinnamon, and so forth. Oil from pumpkin seeds is a common salad oil in Austria and Slovenia (the Styria region) and parts of Hungary.

Consumption of pumpkin seeds and oil from pumpkin seeds were recommended over a long-term period in alternative medical therapy of small disorders of the prostate gland and the urinary bladder. The antioxidative properties of the drugs in pumpkin seeds are also supposed to act as a form of therapy for spontaneously hypertensive rats [12].

Clinical studies of the therapeutic use of pumpkin seed by patients suffering from benign prostatic hyperplasia showed that life quality was improved by 46.1% during the treatment, and the value of International-Prostate-Symptom-score (I-PSS) was decreased by 41.4% [12].

Recently, seeds and oil have also been used to promote HIV/AIDS wellness, due to the identification of high value omega-3-fatty acids in the seeds [13]. Oil has a beneficiary effect on hypertrophy, because it can influence hormone regulation. Seeds, pumpkin oil, and oil cakes are also rich in vitamins.

## 5.1.9   Processing

In producing oil from seeds, traditional "hot" and new "cold" pressing procedures are used. In the traditional pressing method, the pumpkin seeds were milled in a stone mill. After milling and the homogenization of milled seeds with water and salt (cca. 50 kg seeds, 6 l water, 250 g salt), the milled seeds were heated in a pan to a temperature of 100 to 130°C, which led to the formation of the dark green to red ochre colored oil and the typical spicy, nutty aroma.

Pyrazines were originally credited with contributing significantly to the aroma [14, 15], but several volatile [16] and non-volatile [17] compounds have been undertaken to elucidate the temperature levels in order to achieve the typical aroma of the product.

# 5.2   Camelina

## 5.2.1   Introduction

Camelina, also called conio, gold of pleasure, and falseflax (*Camelina sativa* [L.] Crantz), was already being grown during the Iron Age, according to some, but was never mentioned as an oilseed used during medieval times. Later it came to represent an important oilseed in Europe. It remained in Spain, France, and Belgium as a relic for many years. After World War II, Austria, Belgium, the Netherlands, North America, and Russia began planned production of camelina as oilseed. Production of oil known as "totrovo olje" was also revived in the Carinthia region of Slovenia [18].

## 5.2.2   Utilization, nutritional and health value

The oil content of camelina seed ranges from 30 to 44% in dry seed matter with unique composition. The main product from the crop is seed containing highly unsaturated oil with about 15% linoleic acid (18: 2n − 6) and about 38% α-linoleic acid (18: 3n − 3). From a nutritional point of view, both fatty acids are essential, and α-linolenic acid represents OMEGA-3 fatty acids. The oil is suitable for culinary application and technical exploitation, e.g., in varnishes, cosmetics, and so forth. [19]. The fatty acid profile of camelina oils in precentages, according to Budin et al. [20], was: oleic (14.1 to 19.5), linoleic (18.8 to 24.0), linolenic (27.0 to 34.7), eicosenoic (12.0 to 14.9), erucic (0.0 to 4.0), and all others (11.8 to 17.4). Among saturated acids, stearic and palmitic acid prevail. Cholesterol content in camelina oil is several times higher than other plant oils [21] and is recommended for consumption with cold dishes [22]. The oil is rich in vitamin E (α−, β−, ϖ−, δ−tokoferol), an antioxidant of unsaturated fatty acids. Variation of seed quality is partially contributed to genotype but mainly to climatic and soil conditions where camelina is produced. [20]. The raw protein content of the camelina seed is approximately 42.5%. Essential amino acid composition is as follows: 8.15%

of arginine, 5.44% of glycine, 3.96% of isoleucine, 6.63% of leucine, 4.95% of lysine, 4.19% of phenylalanine, 5.09% of proline, 4.25% of threonine, 5.42% of valine, 1.72% of methionine, and 2.12% of cystine (A.A. g 16 g-1 N) [23]. The oil contains less than 4% of otherwise harmful erucic acid, and its share of glucosinolates is lower [24, 25] than with other oil *Brassicaceae*, though the presence of sterols is possible.

In traditional medicine, this somewhat hot oil was used for stomachaches and burns. It was also used as a spice and in different forms of industry. The presence of unsaturated fatty acids confirms a theory of healing properties of this oil and indicates possibilities of its use in ulcer, cardiovascular, autoimmune, carcinogenic, skin disease, and diabetes therapies.

Today, camelina is proposed as an appropriate cover crop in systems where soil is cultivated without the use of a plow, and it is suitable for natural agriculture due to its weed competitiveness. It is also a low input plant for green manure [26]. Further engine trials are needed, however, before the use of camelina ester as an undiluted vehicle fuel can be recommended [27].

## 5.2.3   Botany and genotypes

Camelina (*Camelina sativa* [L.] Crantz, syn.: *Myagrum sativum* L., *Alyssum sativum* Scop.) belongs to the family *Cruciferae* (*Brassicaceae*), the order *Capparales*, the superorder *Dillenianae*, the subclass *Dilleniidae*, and the class *Magnoliatae* (dicotyledonous plants). There is no firm taxonomic classification [28].

Camelina (*Camelina sativa* [L.] Crantz) and another three species of wild camelina: Camelina alyssum (*Camelina alyssum* [Mill.] Thell.), littlepod false flax (*Camelina microcarpa* Andrz. ex DC.), and graceful false flax (*Camelina rumelica* Velen) are grown in Slovenia. They differ in size and shape of seeds.

The original camelina plant is wild camelina, which often grew as a weed near fields and pathways. In German-speaking countries, it was called "false flax" or "gold of pleasure." Fertile cultivars include Iwan (Saatbau Linz, Reichersberg, Austria), CA 13X-17, CA 13X-1S-9, and CA 13X-2S-96. Successful selection work continues in Brunschweig, St. Petersburg, and at the Winnipeg Research Centre, Agriculture and Research Centre, Canada [18].

## 5.2.4   Morphology

The camelina plant is very weak, with a well-branched main root. The stem growing from the rosette of leaves is straight, reaches 0.3 to 1.0 m tall, and is slightly hairy. The height of 32 new cultivars ranged from 64.2 to 78.3 cm. The stem and leaves are very similar to flay; the inflorescence is yellow and has a grape-like shape. The form of the flowers is very similar to flowers of other *Brassicaceae*. Although it belongs to the family of *Brassicaceae*, camelina mostly self-pollinates.

*Figure 5.7* Fruits of camelina.

The fruit is a pod, containing very small seeds of reddish color (Figure 5.7). The weight of 1000 camelina seeds is between 0.7 and 1.0 g, according to older data; new cultivars reach 1.19 to 1.68 g of 1000 seed weight [18].

## 5.2.5   Production

The growing period of camelina is 90 to 100 days. It grows practically everywhere, including high above sea level, in dry and humid climates, and in different soil textures. Camelina can grow in almost every soil and climate.

Despite this general claim, researchers tested 28 genetically different plants with 10 cultivars at two experiment sites in Austria and concluded that production of seeds and oil depend on year and location. Higher rainfall results in high oil content. In extremely favorable conditions, when water was not a limiting factor, the produce achieved was 3250 kg seeds ha$^{-1}$ but the share of oil was lower [29*]. An important low-input feature of camelina is its high level of resistance against plant diseases, which may partly be due

* Wollmann, J. et al., Improvement of *Camelina sativa*, an underexploited oilseed, in *Progress in New Crops*, Janick, J., Ed., ASHS Press, Aleksandria, 1996, 357; with permission from ASHS Press.

to the production of antimicrobially efficient phytoalexins. In Central European countries such as Austria, *Peronospora camelinae* is the only disease of camelina that has been found repeatedly, whereas other diseases and pests have been observed only occasionally [30]. Flea beetle (*Phyllotreta cruciferae*) may also be observed on camelina, although due to its strong growth, control is not usually required. The crop is resistant to blackleg (*Leptosphaeria maculans*), which may be a problem with oilseed rape. It is also very resistant to *Alternaria brassicae*.

Camelina is sown as a rotation crop following basic fall cultivation, after cereals or arable crops. Recommended interrow spacing is 20 to 25 cm, and 6 to 10 kg of seed ha$^{-1}$ are required for sowing. The suggested sowing date for camelina is early spring; dry conditions in the flowering stage lower the produce, whereas enough moisture at the seed-filling stage increases the share of oil.

Harvesting can be carried out using a conventional combined harvester; the crop stands well up to 6 weeks after maturity, before pods begin to lose seeds. According to Slovenian data acquired, 800 to 1200 kg seed ha$^{-1}$ are produced. Products achieved in Austria in 1993 and 1994 with 31 cultivars (from FAL Institute of Agronomy, Braunschweig, Germany; N.I. Vavilov Institute, St. Petersburg, Russia, as compared to standard cultivar 'Iwan' from Saatbau Linz, Reichersberg, Austria) show that in similar climates with newly created variety, produce range from 1605 to 2392 kg of seed ha$^{-1}$ (average 1939 kg ha$^{-1}$, LSD$_{0.05}$ = 215). The average amount of oil produced was 792 kg ha$^{-1}$ [29]. Under experimental conditions in Denmark, the yield of seed was between 2.6 and 3.3 t dry matter ha$^{-1}$, obtained from summer and winter varieties, respectively [31].

### 5.2.5.1 Utilization

The taste and odor of the cold-pressed oil is much more pleasant than the taste of linseed oil, which rapidly turns bitter. The oil is best used with cold dishes [22].

The crop is used in the pharmaceutical and cosmetic industries and also in paint manufacturing, with its byproducts used in animal feeds. It is also an alternative source to oilseed rape for biodiesel.

### 5.2.5.2 Storage

Storage stability and resistance to oxidation at higher temperature are lower than other edible oils. Cold-stored *Camelina sativa* oil retains its quality over a longer period of time without noticeable oxidative changes [22].

## 5.3 Safflower

### 5.3.1 Introduction

Safflower (*Carthamus tinctorius* L.) is also known as false saffron, bastard saffron, thistle saffron, dyer's saffron, suff, kardiseed, kusum, ghurtum, and kazhirak. It has a long history of a traditional dye (color), oil, and herb crop, and according to the trends of use, it remains an important organic product. Safflower is very recognizable under the name false saffron, because its florets were used as a cheaper substitute — adulterant — for saffron (*Crocus sativus* L.), a bulbous plant that was formerly cultivated widely across the Middle East and Asia for dye obtained from the stigmas. Orange dye obtained from safflower florets was the reason for its long domestication [32,33].

The probable origin of cultivated safflower is within an area bounded by the Mediterranean and the Persian Gulf. The plant was mainly distributed by Spaniards or Asians who settled over the world. It was grown in small patches over the greater parts of tropical Asia, Africa, China, and Russia, and for a few decades a new plant in the U.S., Canada, Europe, and Australia. Documented reports cited by Weiss [34] show that safflower was positively identified as a dye plant used in Egypt 4000 years ago; additionally, safflower flowers were found with Egyptian mummies dating back to 1600 BC, flowers produced for dye were used to color cheese and other foods in the first century AD, the dye was used in medicine and as color for food in China around 200–300 AD, and in the first millennium AD it held an important role among Arabs. They believed that safflower had alexipharmic and dia-phoretic properties; thus, it was a source of dye for carpet weavers. In India, it is commonly used as edible oil, and florets are added to rice or bread to give them an orange color.

## 5.3.2   *Botany*

Safflower (*Carthamus tinctorius*) is an annual crop from the family *Asteraceae* (syn. *Compositae*). There are about 25 species in the genus *Carthamus*. Closely related to safflower, according to agronomic importance, are *C. oxyacantha* Bieb. and *C. lanatus* L., originally produced in India and Afghanistan, and in India, respectively [34]. The genotypes of safflower vary according to their domestication and cross breeding, and also according to the oil quality. They are many new developing cultivars. The cultivars with high linoleic acid are Turk-4, BCT 0961, BCT 0963 [35], Morlin, Finch, S-208, S-541; those with high oleic acid are Saf-1 [35], Montola 2000, Montola 2001, S-317, S-518, and so forth. Oleic and linoleic safflower varieties should not be mixed or grown within one mile of each other. In Europe, suggested cultivars are Saffire, CW74, and A24 from California. Spread cultivars also include Alameda, Alarosa, Gila, Saffola, Sepasa, Demetra, Tomejil, Belisario, Sironaria, Bonello, Roberto [33], and so forth, from different breeding programs.

The fleshy taproots with their numerous thin horizontal laterals are a very important characteristic of safflower; they frequently penetrate a depth of 2.5 m, which results in nutrient uptake from deeper layer than many other crops. The plant grows 35 to 200 cm high, is highly branched and thistly, and mainly produces deep yellow to red-orange flowers. Plant height is a genotypic characteristic, affected by climate and provided growth conditions

*Figure 5.8* Branch of safflower.

by cultural practice. The main stem branches about 17 cm to form secondary branches, which themselves branch, each branch terminating in a flower head (Figure 5.8). The leaves are usually large (from 5 cm across to 15 cm long), spineless, and deeply serrated on the lower part of stem; on the upper part, the leaves are smaller (about 2 cm across and 5 to 10 cm long) with spines, but the number of spines is also a genotype characteristic. The number of branches and flowers heads varies with the plant genotype, environmental conditions per plant reducing with the increase of the plant population and/or the decrease of interrow spacing. Usually 3 to 30 flower heads will exist per plant, with a maximum of 180, if the plant has the appropriate conditions for growth.

The inflorescence consists of numerous florets collected together on a circular receptacle. The receptacle is surrounded by several layers of conical bracts, with a small apical opening tube allowing the flowers to protrude; the outer ring of the bracts is slightly spined. The achene is similar to that of the rectangular sunflower seed, only more fibrous and with a thicker seed coat. The testa is cream or off-white in color in many genotypes, but grey and mottled types also occur, due to melanin of the outer schlerenchyma layers resulting from crosses to produce thin-hulled seed. Dark testa indicates crossing with wild species. The seeds are 5 to 9 mm long, and about half as wide. The 1000-seed weight is ranges from 26 to 40 g, and may sometimes reach 100 g. The proportion of hull can vary from 30 to over 60% in commercial hybrids and varieties, though it is less than 45% of the total weight of the seed.

### 5.3.3 Ecology

Safflower is a day-neutral crop with good adaptation of genotypes to specific photoperiods. High yields were obtained in long day circumstances;

sometimes it requires 14-hour days to initiate flowering. The temperature is considered more important than the day length, however. Safflower plant seedlings are susceptible to the frost, which varies between −7.5 and −12°C (sometimes extending to −14°C) for specific tolerant genotypes, and they are susceptible to high temperatures at the flowering stage. The appropriate temperature for flowering is between 22 and 33°C, but under irrigated conditions a temperature of 43°C has no effect on yielding. A heat-resistance profile was obtained by oxygen regulation of the microsomal oleate desaturase ($FAD_2$). The data [36] indicate higher thermal stability and lower dependence on oxygen availability of the safflower $FAD_2$ enzyme than sunflower. This data could help explain why the linoleate content of safflower seeds is more independent of growth temperature than that of sunflower seeds.

High temperatures combined with high humidity can reduce seed yield. The growing period of 130 to 160 days should be frost free, which makes safflower production suitable in all maize-producing regions.

In dry conditions, translocation of preanthesis assimilates to the seed is of great importance for seed growth, because hot and dry conditions during the seed-filling period diminish photosynthesis and crop nitrogen uptake. The contribution of preanthesis accumulated reserves to seed weight ranged from 64.7 to 92.2%, indicating the importance of preanthesis storage of assimilates for attaining high safflower yield [37].

Safflower is a viable alternative for use in rotations where saline soils and irrigation water limit production of nontolerant crops. Safflower tolerated greater levels of salinity than previously reported [38].

## 5.3.4   Possibilities for organic production

Safflower can be planted in organic crop rotations in temperate climate conditions. It can be grown on different soil types, but well-drained, medium-deep sandy loam soils with neutral reaction are recommended for higher yields. Soils with good drainage are important in a humid climate, but in dry areas irrigation is essential although safflower is known as a crop that is drought resistant and appropriate for semiarid condition. Because of susceptibility to root rot (*Phytophtora drechsleri*) and foliar diseases, reserves of water should be added to the soil before planting, or subirrigation or furrow irrigation should be utilized.

Safflower has low demands for plant nutrition, except for high-yielding cultivars proper for irrigated production. Fertilizing is not necessary in organic farming based on broad crop rotation (often with legumes) or the use of animal manure fertilization on fertile soils. On such soils (e.g., well-fertilized barley fields as compared to unfertilized hay crops) in nonirrigated conditions, nitrogen fertilizing increased the yield only 7 to 10% and irrigation was not very effective, with only a 7 to 10% increase of safflower grain yield in fertilized soils [34]. In some cases (e.g., limited growing season), nitrogen top-dressing may even reduce yield.

Deep-rooted crops like safflower used in rotation can improve the overall water and nitrogen use efficiencies of cropping systems and can help minimize nitrate leaching to groundwater. Bassil et al. [39] suggest that nitrogen fertilization applied to safflower could be reduced or even eliminated following crops previously fertilized at high levels. Residual nitrogen should be accounted for in growers' management programs.

Deep-rooted crops can use larger amounts of phosphorus from the soil. If necessary, small amounts of phosphorus as animal manure or rock phosphate may be used in organic farming practice. Potassium fertilizers are unnecessary for safflower except in soils with severe potassium deficiencies.

Safflower can be sown everywhere when the FAO 400 group (mid early hybrids in Corn Belt) of maize can reach maturity; the optimal sowing date is equal for both maize and safflower. Seedbed temperature should be between 12 and 20°C for quick emergence. Seedlings may require up to 30, 21, 12, and 7 days if the soil temperatures at germination and emergence stages are 4–5, 8, 12, and 21°C, respectively. Soil temperature and adequate soil moisture at sowing are the main factors influencing emergence. Late planting dates generally result in shorter plants, decreased branching, and lower seed and oil yields. Artificial drying of the seed is also necessary, if rainy periods appear during harvest time. If safflower is sown in autumn, germination may take place within a wide temperature range, which can increase the time for emergence up to 2 months.

No specialized equipment is necessary to sow safflower; all traditional seed drills for wheat, sunflower, bean, or maize are suitable. Safflower is usually planted in 15- to 18-cm row spacing. Interrow spacing (40, 45, 70, or 90 cm) should be adapted to more recent possibilities for mechanized interrow techniques, mainly mechanical hoeing and weeding. Narrow rows are best for competing with weeds and usually result in more uniform stands that mature earlier. Greater row spacing increases air movement and penetration of sunlight into the crop canopy that may reduce leaf disease incidence but can favor weed competition.

Planting depths of 2.5 to 4.0 cm are optimum. Seed rates for a sole crop in a large field scale vary from 14.0 kg to 60 kg seeds ha$^{-1}$, depending on quality of drilling machines, genotypes, and growth conditions. The suggested final plant population varies between 60 and 90 plants m$^{-2}$.

In organic cultivation practice, weeds may be problematic because mechanical weeding of young plants is difficult. Besides minimizing weed problems with good crop rotation, all possible measures to decrease weed population have to be taken before planting, including furrow leveling, presowing soil preparation, blind sowing, and light harrowing before emergence, which effectively kills small weeds. Although light damage may be caused, rotary hoes and finger weeders can be used while plants are still small. In later stages, only comb harrow can be used during the middle of the day. Once the plants reach 15 cm of height, interrow cultivation can be used and in-row-only hand weeding is possible. Carefully done hand weeding often produces the highest yield. In fields where weeds are a problem at harvest time, the final cultivation has to be delayed as long as possible.

Safflower is usually grown after cereals or fallow in crop rotation. As it leaves little crop residue, fields may be susceptible to wind and water erosion after safflower rotation. Soil loss can be reduced by growing safflower in strips with winter wheat or stubble fallow. Proper crop rotation is especially important for safflower due to its susceptibility to stem rot (*Sclerotinia sclerotiorum*), and it should not be grown in monoculture or in close rotation to other susceptible crops such as dry beans, lentils, soybeans, mustard, sunflower, canola, or rapeseed. Besides crop rotation, no specific measures against this disease currently exist, although following certain hygiene steps on the fields (including the collection and destruction of infected plants, deep plowing, and clean weeding) can help prevent the spread of disease. Some modern organic farmers will spray soil and incorporate microorganisms against *Sclerotina* problems. However, plants are being developed that are resistant to diseases; for example, the progeny of cross 86-93-36A x 237550 showed an immune reaction to most of the analyzed biotypes in the study of resistant progenies on *Fusarium oxysporum* f. sp. *carthami* (Foc) [40].

Leaf spot (*Alternaria carthami*) and bacterial blight (*Pseudomonas* spp.) are usually a problem in humid conditions, especially in years with above-normal rainfall or extended periods of high humidity. Irrigation may also cause a problem with root rot (*Phythophtora drechsleri*). Some current genotypes are tolerant of or even resistant to diseases, as well as leaf spot and bacterial blight. In organic agriculture, treatments with copper are also possible. Crop rotation, cultivar selection, and disease-free seed are the most important prevention methods for the above-mentioned diseases and others, including safflower rust (*Puccinia carthami*).

Many cosmopolitan pests can cause potential problems in safflower production; however, safflower can compensate well for insect damage, and economic losses do not usually occur unless the stands are greatly reduced. Seeds and seedlings may be attacked by cutworms (*Agrotis* spp.), and wireworm (*Agriotes* spp.) damage on leaves can cause different aphis (e.g., green peach aphis *Myzus persicae* and *Macrosiphium* spp.), spider mites (*Tetranychus* spp.), and onion thrips (*Thrips tabaci*). Damage may also be incurred on the flowers and fruit by safflower flies (*Acanthiophilus helianthi*), thrips (*Frankliniella occidentalis*), the lygus bug (*Lygus* spp.), and others. In some cases, grasshoppers have also been known to cause damage. Although also some insecticides can be used on the base of plant extracts (pyrethrum, rotenon, azadirachtin), growers must consider the effects of any insect control measures on beneficial insects and bees, which are attracted to safflower during the flowering period and improve the seed set with their pollination activities.

Safflower is ready to harvest when the plant is quite dry and retains no more than 12% moisture in the achene. The plant should ideally be harvested when not brittle and when the moisture content of the seeds measures 8% or less. The crop can be harvested by grain combine with as much as 14%

of moisture in the seeds, but it should then be artificially dried. The speed of the combine and rotation of the reel must be adjusted to the speeds at which optimal harvesting and minimal losses occur. The yielding can range from a few hundred to 2000 kg grain ha$^{-1}$, but this number may be doubled under experimental conditions [33].

### 5.3.5 Utilization

Safflower is a dye (color), oil, and herb crop, as mentioned previously. Depending on the genotype and growth conditions, oil content has risen to between 25 and 45%, and protein is between 12 and 24%. Safflower oil is used by both food producers and industries. The percentages of monounsaturated fatty acid (oleic) and polyunsaturated fatty acid (linoleic) vary, depending on the chosen cultivars. High oleic oil is beneficial for the prevention of coronary artery disease. The percentage of monounsaturates in safflower oil is higher than in olive oil. Oil from oleic cultivars of safflower is used like other oils, as well as for heat-stable cooking oil used to fry such food items as french fries, chips, and other snack items; for food coatings; and in infant food formulations. Safflower oil is also used for producing margarine. High linoleic oil may be used like a drying agent in paints and varnishes, especially for natural-based houses, because of its nonyellowing characteristics. High linoleic safflower oil is also used in human nutrition, but market demands in recent years have moved toward high oleic oils.

Safflower petals contain yellow color cartamin, or exactly three major yellow pigments: hydroxysafflor yellow, safflor yellow B, and precarthamin [41]. Cartamin is a very useful organic dye of orange-red color, especially for coloring organic foods (such as cheeses and sauces), linen, and other natural canvases, including carpets. When used for coloring, however, it is important to know that thermal degradation reactions of three yellow pigments (hydroxysafflor yellow, safflor yellow B, and precarthamin) at temperatures of 70 to 90°C were founded by first-order reaction kinetics at pH 3.0 and 5.0, but not under neutral and alkaline conditions [41].

Within traditional medicine practices in China, Japan, Greece, Turkey, and other countries, safflower is already used as a mild laxative, a pot-herb, a sedative, and an antirheumatic plant. It can also help to mitigate general health difficulties.

Some modern-day compounds, like antioxidants, are determined from safflower. A total of 10 flavonoids, including quercetin-7-O-(6′-O-acetyl)-beta-D-glucopyranoside [42], (2S)-4′,5-dihydroxyl-6,7-di-O-beta-D-glucopyranosyl flavanone, and 6-hydroxykaempferol 6,7-di-O-beta-D-glucopyranoside [43], have been isolated; additionally, seven known flavonoids, including luteolin, quercetin, luteolin 7-O-beta-D-glucopyranoside, luteolin-7-O-(6′-O-acetyl)-beta-D-glucopyranoside, quercetin 7-O-beta-D-glucopyranoside, acacetin 7-O-beta-D-glucuronide, and apigenin-6-C-beta-D-glucopyrano syl-8-C-beta-D-glucopyranoside, have been isolated from the leaves of the safflower. Among these flavonoids, luteolin-acetyl-glucoside and quercetin-acetyl-glucoside showed potent and oxidative activities against 2-deoxyribose degradation and lipid peroxidation in rat liver microsomes [42]. Luteolin, quercetin, and their corresponding glycosides also exhibited strong oxidative activity, while acacetin glucuronide and apigenin-6,8-di-C-glucoside were relatively less active [44].

Phenolic compounds like quercetin, luteolin, apigenin, isorhamnetin, umbelliferone, and daphnoretin have been determined in common safflower, and acacetin was determined in its flowers [44]. The data also find that herbal extracts containing ginseng and safflower can be a useful anticancer compound against breast cancer [45].

## 5.4   White mustard

### 5.4.1   Introduction

White mustard (*Sinapis alba* L., Syn.: *Brassica alba* [L.] Boiss., *Raphanus albus* Crantz) most likely originates from Asia. It was used 2000 years ago for medical properties and as an important spice. In larger quantities, it is produced in Canada; it also spread throughout Europe, especially to the Czech Republic, The Netherlands, Sweden, Germany, and France, the country especially famous for mustard. In all of these countries, and also in Slovenia, the processing of mustard based on imported inorganic seeds has been preserved. White mustard can be also a late stubble crop for animal feed and a beneficent catch crop [46].

### 5.4.2   Utilization

Grains contain 20.0 to 36.3% raw oil, which can be used as a fuel lubricant, especially the grains of cultivars with high eruca acid content (not appropriate for consumption). Mustard seed contains 26.4 to 32.9% raw proteins and higher quantities of glucosides that give the seed a sharp, hot taste. In some places, emerged mustard seedlings are used for green salad, and the seed is added to salads or vinegar for aroma.

Golden yellow oil is produced from mustard seed by cold pressing and can be used for many purposes. Marc is fed to sheep or transformed to the spice known as mustard. Seeds are also added to vegetable preserves, salads, or wild game sauces.

Besides the aforementioned proteins, the seed also contains 20.8% nonnitrogen extract and 28.2% of joint carbohydrates, 5.2 to 10.3% fibers, and 4.1 to 4.5% ash. In 100 g of seed, there are 410 mg of Ca, 613 mg of P, 20.9 mg of Fe, 630 mg of β-carotene equivalent, 0.40 mg of thiamine, 0.31 mg of riboflavin, 7.3 mg of niacin, and no ascorbic acid. Among glucosides, sinalbin is the best known (1.5 to 2.5% of grain content), used in hydrolysis (mysorin enzyme fermentation) for essential oils and sinalbin sulphate production. Both essential oils and sinalbin sulphate have been used to heal chronic enteritis, headaches, and more. In folk medicine, white mustard was used internally and externally; it was thought to heal numerous diseases, even cancer, bronchitis, and other respiratory organ conditions, to be a strong diuretic, and to reduce local and rheumatic pains.

White mustard was mentioned in 1699 (in John Evelyn's *Acetaria*) as a reviver of the spirit and improver of memory. The use of white mustard seed for health was also described in 1834 by Sinclair and Siclair [47]. The powerful effects of this plant can be attributed to tranquilizers, sedatives, and narcotics that are found in the grain. Some people have allergic reactions to mustard seeds because of allergens based on 2S proteins (protein napins – Sin $a_1$). Sin $a_1$ consists of two polypeptide ties with 38 and 88 amino acids, connected by disulphide bridges with alpha structure; as a result, Sin $a_1$ displays proteolytic and heat resistance. The same allergen was also found in oil mustard.

White mustard is a melliferous plant. Nectar secretion starts at the loose bud and peaks during anther dehiscence. The average amount of nectar secreted by 100 flowers of white mustard is 134 mg, and the nectar secretes 24.9 mg of sugars. The estimated sugar productivity and total honey yield of the crop is between 71.2 kg [48] and 120 kg ha$^{-1}$.

## 5.4.3   Botany and genotypes

For white or yellow mustard, *Sinapis alba* is produced (*Sinapis alba* L., syn.: *Brassica hirta* Moench, *Brassica alba* [L.] Rabenh.). Seeds of *Brasica juncea* L. and *Brasica nigra* L. are also added to mustards.

The cultivars that can be grown for grain and oil are as follows: Budakalaszi sarga, Emergo, Albatros, Arda, Asta (R), Condor (R), Emergo (R), Litember, Maxi (R), Salvo (R), Serval (R), and Signal. Cultivars that prevent development of beet nematodes (R) can be produced for grain in early spring. The cultivars Mirly and Serval are suggested for sowing as a stubble-catch crop or a crop for green manure, i.e., for incorporating organic matter into the soil.

The roots of white mustard are less developed than rapeseed. The steams are rude and reach between 0.7 and 1.5 m in height, with an average of 0.8 m. Leaves are 4 to 6 mm long and hairy, and inflorescence is a short, clustered raceme. The flowers are composed of four sepals and four 7 to 10-mm long petals, six stamens, and one pistil. It can be pollinated by wind or insects and is suitable for bee-pasture. Bees are more important pollinators than wind.

The fruit is a pointed and pilose pod, measuring 2.5 to 4.0 cm in length (Figure 5.9). The seeds are yellow-white, and the 1000 g weight of the seeds ranges between 3 to 6 g.

## 5.4.4   Growth

The growth period of white mustard is approximately 90 to 105 days. Photoperiodically it is considered to be an explicitly long-day plant. Consequently, it is important that it is not sown too late in spring because the

*Figure 5.9* Inflorescence of white mustard.

influence of day length on the flowering stage causes plats to remain too small. Stubble-crop sowing for green manure worsens vegetative growth and development due to short days. Plants are sensitive to severe spring frozen fog, and young plants freeze at –4°C. Mustard prefers milder and warm climates with moderate humidity.

## 5.4.5  Production

White mustard is not too demanding regarding soil type, but it prefers sandy-loam and marl soil. It should primarily be grown in pH-neutral soils, because the plant is quite sensitive to soil acidity.

It can be grown after arable crops fertilized by animal manure in crop rotation. At the very least, a 4-year interval should take place between the next sowing or between sowing the *Cruciferaceae* or beets. It serves as a good previous crop due to its ability to cover soil in the spring and summer months in short growth periods. White mustard is sown as a catch crop to limit nitrate leaching during winter before a spring-sown crop. Fields covered with white mustard sustained higher net mineralized nitrogen levels than the fallow control after incorporation [49, 50] or the fields without cover crops. The fate of nitrogen from N-15-labeled white mustard like catch crop is as follows: (i) the initial decline in N-15-labeled organic N, after incorporation of the material in early December containing 2.6% N; (ii) After 33 months of decomposition, 34% of the mustard N-15 was recovered in organic residues in the topsoil [50, 51]. White mustard is a good precrop for most stubble crops (crops, fodder, and vegetables), whereas it is not recommendable to sow it prior to seed beet and *Brassicaceae* unless the cultivar displays nematode resistance to beet nematodes. White mustard's inclusion in crop rotation should also reduce the buildup of inoculum of soil born fungal pathogens like *Pythium* spp., *Aphanomyces cochlioides*, and *Rhizoctonia solani* and offers a tactic to relieve the problem in the longer term [52].

White mustard is a medium input plant, according to nutrient needs. The needs should be calculated as: 50 to 80 kg N ha$^{-1}$, 40 to 60 kg P$_2$O$_5$ ha$^{-1}$, and 70 to 90 kg K$_2$O ha$^{-1}$. The phosphorus and potassium are plowed in autumn or reduced by half, with the other half added during presowing cultivation. Sowing should take place in early spring. Later sowing results in a lower yield, due to reaction on long-day period (photoperiodism). White mustard is sown at 25 to 30 cm interrow spacing with 10 to 15 kg seed ha$^{-1}$ or, according to older Slovenian data for manual sowing, 20 kg seed ha$^{-1}$. Naturally, the seeding norm depends on the sowing machines. American literature on the subject cites requirements of 4 to 5 kg seed ha$^{-1}$, while Great Britain uses 12 kg in heavy soil and 14 kg seed ha$^{-1}$ in light soil. Emergence observations indicatedthat coarse seedbeds and large quantities of straw are not as important as soil water content, temperature, and sowing depth in emergence [53]. Due to its quick growth, white mustard is hoed only once and thinned if necessary. For sowing *Brassica juncea* L. and *Brassica nigra* L., 10 to 14 kg seed ha$^{-1}$ are needed.

There is a lack of data on economically important losses due to diseases in white mustard. The plant may be attacked by a wide spectrum of pests, from the seedling stage to maturity. Often most damaging to young plants are flea beetles (*Phyllotreta* ssp. and *Psylliodes* spp.) and beetles that attack the reproductive parts, such as pollen beetles (*Meligethes assimilis* Payk.), seed weevils (*Ceutorhyhnchus assimilis* Payk.), sawflys (*Anthalia* ssp.), and so forth. Decisions should be made carefully for organic white mustard growing in regions with damaged biodiversity due to related species in the monoculture or narrow crop rotation. Flea beetles can ward off the young plants and damage it with the intercropping of buckwheat (sowing into the strips can be suggested).

White mustard is harvested when the field turns golden-green and grain is firm enough. Grain does not fall out at harvest because the pods of this crop do not shatter. A yield between 1200 and 1600 kg seed ha$^{-1}$ can be expected.

## 5.4.6 Processing of new organic products

White mustard is a good emulsifier that may be used for processing mustards. Due to a lack of organically grown white mustard, organic mustard is often difficult to find in marketplaces. There are a few ways to produce mustards from fine grinding seeds. The mustards can be mixed with whole seeds, herb spices, and red wine stone, along with the whole or grinding seeds of other crops. Organic edible pastas from sunflower or pumpkins seeds in combination with mustards and other spices have also been produced, mostly on a small scale for vegetarian consumers. The portions of the components depend on individual taste.

Oil processing from white mustard seeds is not very common. The separation of the oil causes problems; the mucilage on the surface of the seeds during aqueous processing acts as an emulsifier. Temperature and time scale should be taken into consideration for more effective processing [54].

### 5.4.7   Production for green salad

White mustard is produced for green salad in greenhouses. It is picked when the first pair of leaves develops and produced at a temperature between 10 and 15°C. Seed is sown on the soil surface with a compact seedbed, moisturized by fine spraying, and covered with a sterile cover for moisture preservation. The cover is removed when plants reach a height of 2.5 to 3.5 cm, which usually occurs after 4 days in spring and 6 to 7 days in winter. Within 2 to 3 days, yellowish leaves turn green and the yield is storied. Produce is packed in small boxes either on its own or mixed with watercress.

## 5.5   (Garden) poppy

### 5.5.1   Introduction

Poppy with edible seeds is also called oil (seed) poppy or garden poppy (*Papaver somniferum* L. ssp. *somniferum* Kadereit); there is a large group of poppy genotypes described. The synonyms of garden poppy described in this book are as follows: *P. officinale* Gmel., *P. somniferum* var. *album* DC, *P. album* Crantz, and *P. somniferum* var. *hortense* Hussenot. Poppy is a plant from the genus *Papaver* with many known species [55]. One of the well-spread weeds is *Papaver rhoeas* L., which has not been recognized as the ancestor of the garden poppy (*Papaver somniferum* L). *Papaver setigerum* was accepted as the first ancestor, which served as a cultivation species in the Mediterranean area and later in other poppy-growing places. Some form of poppy was widely grown in the territories of France, Germany, Hungary, and Spain in 4000–5000 BC [55]; however, the gene center of the poppy is located in middle [56] and western Asia. It is considered one of the oldest cultivated plants and was known as far back as the Stone and Bronze Ages.

Poppy was produced by Greeks and Romans. Especially popular with old Greeks, it later spread to the Arabic world, China, India, and Bulgaria, where it was processed for raw opium using the green plant parts that contain a white and poisonous milky substance. The milky-colored substance consists of 44 identified alkaloids (morphine, codeine, narcotine, papaverine, tebaine, and more) used in the pharmaceutical industry as irreplaceable substances for medicines. The genotypes that contain highly concentrated opiate alkaloids are used for this kind of production, and they are grown in hot areas, mostly tropical climates.

Morphine is the most important alkaloid in the pharmaceutical industry; raw opium contains 10.95 to 13.42% of morphine and 14.90 to 18.22% can be found in dry matter. Morphine is also the most commonly used painkiller in medicine. Previous belief held that mature seeds did not contain opiate alkaloids, but they could be extracted from harvested poppy heads; modern methods, however, provide extraction from the entire plant. It was also believed that poppy seeds from temperate zones did not contain opiates. Recent data show that the content of opiate alkaloids in 28 cultivars of the Hungarian genetic bank Tapioszele varies from 100 to 8600 g k$^{-1}$ [57]. Poppy genotypes of oil (seeded) garden poppy, which are grown in temperate zones, generally contain low amounts of opiates.

Poppy seeds often appear in food products [58]. However, in this book we feel that we can make a clearer separation of cultivars without opiate alkaloids or with low content focusing on these kinds of substances.

## 5.5.2   Morphology [58]

The garden poppy is an annual or biennial plant if sown in autumn. Most parts of the well-developed roots are found in the surface soil layer. In favorable soil conditions, the main root can reach a depth of 1.0 to 1.5 m, but lateral roots are less developed. Roots absorb nutrients very well. Narrow oval and threadlike hypocotyls, which end with even thinner roots, are characteristic for a young plant. The first two true leaves appear simulta-neously and on opposite sides. As a rule, one leaf will be bigger than the other. First leaves are ovate in shape and somewhat pointy at the top; at the base, they continue into leaf stalk. The next leaves are 10 to 30 cm long and appear in rosette after the first pair of leaves. Leaf pedicle appears only with the bottom leaves; the upper leaves are sessile, usually oval and divided. Upper leaves can embrace part of or all of the stem's circumference. Leaves of oil (seeded) poppy have a waxy surface and are thinner than opium poppy.

Plants grown in optimal conditions reach 0.7 to 1.5 m tall and their stems are straight, divided and covered with waxy surfaces. The plant is unbranched or branched only in the upper two-thirds of the stem; each branch forms a flower.

Flowers are large with 2 sepals, 4 petals, approximately 200 stamina, and a pistil. Flowers on the main stem blossom prior to the flowers on side stems (Figure 5.10). Opium poppy flowers are white, white-purple, or purple, while oil (seed) poppies have reddish or bluish flowers. Transitions from purple to bluish color are difficult to determine. Poppy is partially an autogamous species, and its share of allogamy depends on the presence of insects and even on the wind. Flowering is gradual, and main stem flowers open first. Until they open, flower buds are turned downward. When opening, flowers turn to a horizontal position and the stalks become firm. The flowers open in the morning.

The fruit (capsule) can take different shapes (clavate, elipsoidal, ovate, globose, obovate, and so forth). The shape of capsule can be determined by the poppy capsule index ($I_m$), which can help us differentiate cultivars and make selections.

*Figure 5.10* Poppy at the flower formation and flowering stages.

$$I_m = \frac{\text{Poppy capsule width}}{\text{Poppy capsule length}} \times 100$$

With round capsules, $I_m$ ranges from 75 to 125; with oval leaves, $I_m$ ranges from 175 to 270; with wide oval leaves, $I_m$ ranges from from 125 to 175; and with flattened leaves, $I_m$ ranges from 45 to 75.

Thick capsules with diameters exceeding 3.5 cm are desired. The main characteristic of oil (seed) poppy is its ridged shape; opium poppy capsules are smooth. The highest quantities of milky-colored alkaloid substances are found in the green heads of opium poppy. The oil (seeded) poppy capsule is compact and divided into several parts with attached seeds. Ordinary opium poppies have no compartment capsules.

Capsules contain between 500 and 2000 seeds. Mature capsules contain 54 to 60% of seeds and may crack open with some cultivars. Seeds are small and round and can be white, grey, blue, pale yellow, reddish, or black. The 1000-seed weight is between 0.2 and 0.7 g, with a hectoliter weight between 50 to 60 kg hl$^{-1}$. Austrian authors [59] have found that 1000-seed weight can grow to between 0.2 and 0.4 g, or even 0.6 g, depending on production conditions and cultivar.

*Figure 5.11* Poppy at the stage of leaves rosette.

### 5.5.3 Growth stages

The growth stages of oil (seed) poppy last between 100 and 140 days (approximately 125 days on average), while opium poppy requires between 240 and 260 days.

The following poppy growth stages are differentiated: seed germination, plant emergence, stage of first true leaf, rossete stage (Figure 5.11), stage of stem growth, flowering stage, stage of end of flowering, budding stage and fruit (capsule) formation, and stage of maturity.

The emergence stage begins when hypocotyls appear on 10% of the seeded field and ends when 75% of emerged cotyledons take a horizontal position. From this point to the point when 75% of the plants have developed first true leaves, the stage of first true leaf takes place. The rosette stage follows, similar to the rapeseed rosette stage. The stem growth stage begins with main stem growth of 10% of the plants and lasts through the appearance of buds until 10% of flowers open. The flowering stage is represented by 75% of opened flowers. A day or two later, the end of flowering occurs and the budding stage takes place. The beginning of the maturity stage is noted by the permanent form and size of capsules. Capsules are light green and then turn paler; at full maturity, they will turn yellow and grey. Seed is ripe when it detaches from the placenta. At this stage, a distinctive rustle can be heard when shaking a capsule.

### 5.5.4 Production and cultivars

Oil (seed) poppy and opium poppy are spread differently. Opium poppy is optimally produced between 27 and 41° northern latitude and can be grown to 48° northern latitude, though with poorer results. Oil (seed) poppy is a cosmopolitan crop; it is produced to 54° northern latitude, into the southern borders of Sweden and Norway. The highest quantities of oil (seed) poppy are produced in Russia, whereas opium poppy is produced mostly in India. Consequently, opium poppy is a plant of subtropical continental climate. Although oil (seed) poppy thrives in the areas of opium poppy, it is not produced in the same area due to the higher profitability of opium poppy. Oil (seed) poppy is therefore produced in northern areas for seed [58].

The number of poppy cultivars is extensive. For oil (seed) poppy production in the temperate European zone, the following blue cultivars are recommended besides Marrianne: Parmo, Rosemarie, Przemko, Niebieski, Neuga, and Libra. Among grey seed cultivars, the recommended ones are: Florian, Edelveiss, and Edelrot; a recommended white seed cultivar is Gusentaler. Of the desired early cultivars, the blue seed cultivar Zeno is recommended [33, 58, 59] together with declared opium and alkaloid-free seed poppy cultivars like Sujata [60, 61].

### 5.5.5 Growth characteristics and organic cultural practice

Temperature sum necessary for growth and development of poppy is from 2200 to 2800°C for growing period. Poppy can emerge at minimum temperatures between 2 and 3°C. Optimal field emergence is guaranteed at 10–11°C, and laboratory optimum is reached at 30°C. During the initial development stage, temperature demands are low and increase with growth; poppy is also resistant to low temperatures at this point. Notable temperature fluctuations will result in uneven growth and development. Young plants can survive short-term frosts between –10 and –12°C. Winter poppy can freeze at a sustained temperature of –6°C without snow cover. When plants are covered with snow, they can survive temperatures as low as –15 or even –20°C. Because young poppy is resistant to low temperatures, it can be sown up to 800 m above sea level.

Poppy demands regarding moisture are very high; its transpiration coefficient is approximately 500. The most water needs to be available during the quick growth stage, the flowering stage, and the grain filling stage. During the growing stage, poppy requires 180 to 200 mm of well-distributed precipitation. Poppy is a long-day plant. In addition to heat and moisture, it is also extremely sensitive to intensity of light. For early poppy, at least 700 sunny hours are needed.

Poppy grows well in structural and deep soil. The most appropriate soil pH range is 6.5 to 7.0. Soil intended for poppy production needs a lot of calcium, and according to textures in our conditions, the most appropriate soil is sandy loam.

#### 5.5.5.1 Oil (seed) poppy production techniques [58]

Any crop that precedes poppy growth should leave the soil weed-free, due to the poppy's slow growth in the early development stage. The best precrops are root crops, tuber crops, winter cereal, legumes, and sunflowers. In Slovenia, poppy is produced as intercropped plant with beet or other hoe crops; it is a part of amateur production and sown only by natives. Poppy is the most suitable precrop for winter cereals. Monoculture is not suitable for poppy; it should be sown only every fifth or sixth year on the same field.

Seedbed for oil (seed) poppy is prepared in the following order: basic plowing to the usual depth and rough field leveling in autumn, with spring leveling if necessary; presowing cultivation with a machine for destroying emerging and emerged weed; and finally flattening with a smooth roller. Due to extremely small seeds that need to be sown at appropriate depths, a leveled field and suitably stable structural soil are very important.

Because of poppy's short vegetation period, sufficient nutrients must be available. Nutrient uptake with production of 1250 kg seed ha$^{-1}$ and aboveground mass is considerably large, according to test results: 102.5 kg N ha$^{-1}$, 190.0 kg $P_2O_5$ ha$^{-1}$, and 113.4 kg $K_2O$ ha$^{-1}$.

In the autumn, 15 to 20 t ha$^{-1}$ of manure should be plowed in soil but without weed seeds. Too much manure and, consequently, too much mineral nitrogen (N min) in the soil may prolong the length of the growing stage or even reduce production. Sometimes manure is more suitable for use on the preceding crop. The phosphorus supply must be noted for (oil) seed poppy; it should be added in allowed amounts during organic production if the soil does not provide sufficient quantities.

It is possible to grow winter (autumn-sown seed) and spring (spring-sown seed) poppy. In temperate zones, where poppy seeds are produced for consumption, spring poppy cultivation is more widespread. Oil (seed) poppy is sown in early spring due to its resistance to cold and slow development. It is sown at the same time as early cereals, or even sooner. For sowing, 3 to 5 kg of seed ha$^{-1}$ may be used; however, some data cites 1.0 to 1.5 kg of seed ha$^{-1}$ as a sufficient amount. In recent recommendations, 0.6 of seed ha$^{-1}$ (at 90% emergence) is mentioned. The correct amount of seed ha$^{-1}$ is based upon the use of more- or less-accurate seeders. When sowing poppy in some locations, 3 to 4 mm coating seeds are also used. Usually, 4.5 of seeds ha$^{-1}$ are sown to achieve plant density of 10 to 20 plants m$^{-2}$; in this case, thinning is not necessary. When using machines with 40 to 60 cm spacing, the optimal plant density is 10 to 13 plants m$^{-2}$.

Sowing in 40 to 50 × 15–20 cm bands is also recommended. German literature on the subject advises sowing with 25 to 35 cm interrow spacing, with a density of 40 to 60 plants m$^{-2}$.

The optimal sowing depth is 0.3 to 0.5 cm, but it is practically impossible to sow at such depth. Therefore, the usual sowing depth is closer to 1.0 to 1.5 cm. If sowing goes deeper than 2 cm, week seedlings cannot reach the surface. In cases of uneven sowing at more than 2 cm depth, the quantity of seeds should be increased.

Poppy emerges in 4 to 6 weeks; therefore, it is advisable to sow poppy with marking plants such as lettuce or other appropriate plants. Poppy can be rolled over by a Cambridge roller 10 days after sowing due to its late emergence. Poppy growth is very slow at the early stages, and small plants are unable to compete with weeds. When sowing in wide rows, some begin to hoe interrow spacing 3 to 4 days after emergence. Seedlings also have to be thinned at the stage of 3 to 4 true leaves and when plants reach the height of 10 to 15 cm.

Poppy is first thinned to 5 cm spacing and later to 15 cm. It can be thinned only once; in this case, plants are too small or bigger plants are unevenly developed due to mutual competition. It is advisable to hoe the crop for the second time before the first thinning and consequently to destroy some weeds. Third hoeing takes place after 2 weeks. When using mechanical methods, weeding may be needed three to five times during the vegetation period or even until the budding-flowering stage. While thinning in organic farming, compost should be added.

To compensate for the lack of moisture, a maximum of 30 mm of water should be added for irrigation. Peronospora agents appear in poppy production (*Peronospora arborescens*), as well as some fungi (like *Fusarium* spp., *Erwinia* spp. and *Helminthosporium papaveris*), pests weevils (*Ceutorhynchus maculaalba*), poppy fly (*Dasyneura papaveris*), and poppy gnat (*Perrisia papaveris*) [55, 58].

## 5.5.6   Harvesting and yield

Oil (seed) poppy reaches full maturity from the middle of August to the middle of September. For seeds, harvesting is done at the stage of full maturity when poppy capsules turn dark yellow. If poppies do not crack, waiting at the full maturity stage for the appropriate harvesting moisture is advisable. Mature capsules contain seeds that rustle when shaken. When harvesting cracking poppy with seeds falling out of capsule cells, it is best to wait for 75% of capsules to mature. Cultivars with cracking capsules are sown manually. Plants are tied into sheaves to dry, and adapted threshing machines are used to thresh. When combining, capsules should be reaped with the smaller part of the stem.

Poppy seed should be very dry; the moisture cannot exceed 10%. Seed that is too moist can spoil quickly. Due to high oil content, seed can turn rancid, especially when stored at high temperatures. Yield usually varies between 500 and 700 kg of poppy seed ha$^{-1}$, although in favorable conditions, higher yield is possible (800, 1200, or even 2000 kg of seed ha$^{-1}$).

## 5.5.7   Use, nutritional composition, and research results about opiates in the seeds

Poppy seed is widely used as an additive to bread and cakes (e.g., the traditional food in Slovenia, Prekmurska gibanica). Often poppy is also produced as an oil crop because of the 50% oil content in its seeds. Poppy oil is of high quality; it contains approximately 70% of oleic acid and 30% of other acids. Cold-pressed oil should be used, as in organic farming. With cold pressing, all of the oil is not pressed and 10 to 50% of seed oil can remain in oil cakes. Oil cakes contain around 40% of raw proteins similar to quality fodder, 20% of carbohydrates, and 13% of raw fibers. Hot extraction procedures give poppy oil a sharp flavor and produce oil suitable only for industrial processing — mostly for paints, soap, and so forth [58].

As stated in the introduction of this chapter, poppy genotypes grown in temperate zones contain low concentrations of opiate alkaloids in the poppy plants. It is confirmed [62, 63] that notable variations in opiate alkaloid concentrations may arise due to variations in the climate, soil composition, seed quality, the year of harvest, and the variety of poppy cultivated. The amount of total alkaloids accumulated in capsules was usually higher under tropical conditions characterized by higher temperature and illumination; this shows that the European cultivars Kék Duna and Reading differ from Indian, Thai, and Afghan ecotypes [64]. Meadway et al. [65] conclude that the country of origin prior to ingestion will influence opiate alkaloid concentration, and that this, in turn, will influence the likelihood of obtaining a positive screening result of alkaloids in urine. The differences in opiate concentration between the cooked, sieved, and untreated seed specimens demonstrated that the method of seed preparation influenced the determined alkaloid concentrations [64–66] and thereby the likelihood of obtaining opiate-positive results. Washing the seeds prior to use has been shown to remove as much as 45.6% free morphine and 48.4% free codeine [66]. However, the universally accepted 300 ng ml$^{-1}$ cutoff limit for opiate assays in urine 6, 24, and 48 hours post-ingestion of poppy products should exceed this value. In Germany, five volunteers ate poppy seed products (50 mg morphine kg$^{-1}$ poppy seeds); all on-site urine tests were positive for opiates and morphine, but all the blood samples were negative (at concentration of 10 ng ml$^{-1}$). Accordingly, penalties based on German law 24a StVG are not likely to cause road users any concerns, should they have consumed poppy seeds [67*].

Such facts are important in order to avoid concern regarding poppy seed consumption. However, consumers may try to choose seeds from genotypes grown in temperate zones and limit the ingestion of poppy seeds among children and people with low body weight, as indicated by Meadway [65], although negative results were also presented [68, 69]. Based on a mutation of opium, the alkaloid-free seed poppy cultivar Sujata was developed and serves as a safe and potential food crop with protein-rich seeds and healthy unsaturated seed oil [60–61].

* Reprinted from *Forensic Sci. Int.*, 143 (2-3), Moeller, M.R., Hammer, K., and Engel, O., Poppy seed consumption and toxicological analysis of blood and urine samples, 183, Copyright 2005, with permission from Elsevier.

# References

1. Bavec, F., Navadna buča, oil pumpkins (*Cucurbita pepo* L. *convar. giromantiina var. oleifera* Pietsch., syn. *convar. citrullina* [L.] Greb. var. *styriaca* Greb.), in *Some of Disregarded and/or New Field Crops* (Slovene language), Univ. of Maribor, Faculty of Agriculture, Maribor, 2000, 21.
2. Teppner, H., *Cucurbita pepo* (*Cucurbitaceae*) — history, seed coat types, thin coated seeds and their genetics, *Phyton–Annales Rei botanicae*, 40 (1), 1, 2000.
3. Bavec, F. and Bavec, M., Evaluation of target value of nitrogen fertilizing on the basis soil N min in Pumpkins (*Cucurbita pepo* L.), presented at 5th Congress of Eu. Soc. Agron. Short Commun., Nitra, Slovakia, 1998, Vol. II, 80.
4. Tagizhed, A., Die Wirkung der Düngung und des Nährstoffaufnahme bei ölkürbis (*C. pepo* L.), Ph.D. thesis, University Giessen, Giessen, 1967.
5. Bavec, F. et al., Seedlings of oil pumpkins as an alternative to seed sowing: Yield and production costs, *Die Bodenkultur*, 53 (1), 39, 2002.
6. Hillebrand, A. et al., Ein hoher Gehalt an Vitamin E und ungesattingen Fettsauren als neues Zuchtziel des Kurbiszuchters, *Ernahrung*, 20, 525, 1996.
7. Idouraine, A. et al., Nutrient constituents from eight lines of naked seed squash (*Cucurbita pepo* L.), *J. Agr. Food Chem.*, 44, 721, 1996.
8. Murkovič, M. et al., Variability of fatty acid content in pumpkin seeds (*Cucurbita pepo* L.), *Eur. Food Res. Technol.*, 203, 216, 1996.
9. Murkovič, M. et al., Variability of vitamin E content in pumpkin seeds (*Cucurbita pepo* L.), *Eur. Food Res. Technol.*, 202, 275, 1996.
10. Wagner, F.S., The health value of Styrian pumpkin-seed oil — science and fiction, *Cucurbit Genet. Coop.*, 23, 122, 2000.
11. Zuhair, H.A., El-Fattah, A.A.A., and El-Sayed, M.I., Pumpkin-seed oil modulates the effect of felodipine and captopril in spontaneouusly hypertensive rats, *Pharmocol. Res.*, 41, 555, 2000.
12. Schiebel-Schlosser, G. and Friederich, M., Phytotheraphy of BPH with pumpkin seeds-multicentric clinical trial, *Z. Phytother.*, 19, 71, 1998.
13. Zimmerman, J.R.D., Optimal nutrition for HIV/AIDS wellness, *J. Am. Diet. Assoc.*, 97 (9), A18, 1997.
14. Nikiforov, A. et al., Zur Bestimmung der dominierenden Geruchskomponenten von steirishem Kürbiskernöil, *Ernährung/Nutr.*, 20, 607, 1996.
15. Buchbauer, G., Boucek, B., and Nikiforov, A., On the aroma of Austrian pumpkin seed oil: Correlation of analytical data with olfactoric characteristics, *Ernährung/Nutr.*, 22, 246, 1998.
16. Siegmund, B. and Murkovi, M., Changes in chemical composition of pumpkin seeds during the roasting process for production of pumpkin seed oil (II: Volatile compounds), *Food Chem.*, 84 (3), 367, 2004.
17. Murkovič, M. et al., Changes in chemical composition of pumpkin seeds during the roasting process for production of pumpkin seed oil (I: Non-volatile compounds), *Food Chem.*, 84 (3), 359, 2004.
18. Bavec, F., *Camelina sativa* (L.) Crantz., in *Nekatere Zapostavljene in/ali Nove Poljščine (Some of Disregarded and/or New Field Crops)*, Univerza v Mariboru, Fakulteta za kmetijstvo, Maribor, 2000, 105.
19. Zubr, J., Qualitative variation of *Camelina sativa* seed from different locations, *Ind. Crops and Prod.*, 17, 161, 2003.

20. Budin, J.T., Breene, W.M., and Putman, D.H., Some compositional properties of camelina (*Camelina sativa* L. Crantz) seeds and oils, *J. Am. Oil Chem. Soc.*, 72 (3), 309, 1995.

21. Shukla, V.K.S., Dutta, P.C., and Artz, W.E., Camelina oil and its unusual choresterol content, *J. Am. Oil Chem. Soc.*, 79, 965, 2002.

22. Matthaus, B., *Camelina sativa* — revival of an old vegetable oil? Ernahrung-Umschau, 51 (1), 12, 2004.

23. Zubr, J., Dietary fatty acids and amino acids of *Camelina sativa* seed, *J. Food Qual.*, 26, 451, 2003b.

24. Lange, R. et al., Glucosinolates in lindseed dodder, *Fett. Wiss. Technol.*, 97 (4), 146, 1995.

25. Drev, I., Vpliv medvrstnega razmika in odmerkov dušika na rast in pridelek navadnega rička (*Camelina sativa* L.), Graduate thesis, University of Maribor, 2001.

26. Putnam, D.H. et al., Camelina: A promising low-input oilseed plants, in *New Crops*, Janick, J., Ed., Wiley, New York, 1993, 314.

27. Frohlich, A. and Rice, B., Evaluation of *Camelina sativa* oil as a feedstock for biodiesel production, *Ind. Crop. Prod.*, 21 (1), 25, 2005.

28. Onylagha, J. et al., Leaf flavanoids of the cruciferous species, *Camelina sativa, Crambe* ssp., *Thlapsi arvense* and several other gemera of the family *Brassicaceae*, *Biochem. Syst. Ecol.*, 31 (11), 1309, 2003.

29. Wollmann, J. et al., Improvement of *Camelina sativa*, an underexploited oilseed, in *Progress in New Crops*, Janick, J., Ed., ASHS Press, Alexandria, VA, 1996, 357.

30. Vollmann, J., Steinkellner, S., and Glauninger, J., Variation in resistance of camelina (*Camelina sativa* L.) to downy mildew (*Peronospora camelinae* Gaum.), *J. Phytopath-Phytopathologische Zeitschrift*, 149 (3–4),129, 2001.

31. Zubr, J., Oil-seed crop, *Camelina sativa*, *Ind. Crops*, 6 (2),113, 1997.

32. Bavec, F., *Carthamus tinctorius*, in *Nekatere Zapostavljene in /ali Nove Poljščine (Some of Disregarded and/or New Field Crops)*, Univerza v Mariboru, Fakulteta za kmetijstvo, Maribor, 2000, 176.

33. Buchgraber et al., *Produktions — Nischen im Pflanzenbau*, Leopold Stocker Verlag, Graz, 1997, 13.

34. Weiss, E.A., *Oilseed Crops*, Longhamn Group Ltd., London, 1983.

35. Mirza, S.H. et al., Variability in oil, protein and fatty acid in safflower (*Carthamus tinctorius* L.) genotypes, *Bangladesh J. of Bot.*, 27 (1), 31, 1998.

36. Esteban, A.B. et al., Growth temperature control of the linoleic acid content in safflower (*Carthamus tinctorius*) seed oil, *J. Agric. Food Chem.*, 52 (2), 332, 2004.

37. Koutroubas, S.D., Papakosta, D.K., and Doitsinis, A., Cultivar and seasonal effects on the contribution of pre-anthesis assimilates to safflower yield, *Field Crops Res.*, 90 (2–3), 263, 2004.

38. Bassil, E.S. and Kaffka, S.R., Response of safflower (*Carthamus tinctorius* L.) to saline soils and irrigation I. Consumptive water use, *Agric. Water Manage.*, 54 (1), 67, 2000.

39. Bassil, E.S., Kaffka, S.R., and Hutmacher, R.A., Response of safflower (*Carthamus tinctorius* L.) to residual soil N following cotton (*Gossypium* spp.) in rotation in the San Joaquin Valley of California, *J. Agric. Sci.*, 138, 395, 2002.

40. Sastry, R.K. and Chattopadhyay, C., Development of *Fusarium* wilt-resistant genotypes in safflower (*Carthamus tinctorius*), *Eur. J. Plant Pathol.*, 109 (2), 147, 2003.
41. Yoon, J.M. at al., Thermal stability of the pigments hydroxysafflor yellow A, safflor yellow B, and precarthamin from safflower (*Carthamus tinctorius*), *J. Food Sci.*, 68 (3), 839, 2003.
42. Lee, J.Y. et al., Antioxidative flavonoids from leaves of *Carthamus tinctorius*, *Archives of Pharmacal Research*, 25 (3), 313, 2002.
43. Li, F., He, Z.S., and Ye, Y., Flavonoids from *Carthamus tinctorius*, *Chin. J. Chem.*, 20 (7), 699, 2002.
44. Suleimanov, T.A., Phenolic compounds from *Carthamus tinctorius*, *Chem. Nat. Compd.*, 40 (1), 13, 2004.
45. Loo, W.T.Y., Cheung, M.N.B., and Chow, L.W.C., The inhibitory effect of a herbal formula comprising ginseng and *Carthamus tinctorius* on breast cancer, *Life Sci.*, 76 (2), 191, 2004.
46. Bavec, F., White mustard (*Sinapis alba* L.), in *Nekatere Zapostavljene in /ali Nove Poljščine (Some of Disregarded and/or New Field Crops)*, Univ. Maribor, Faculty of Agriculture, Maribor, 2000, 33.
47. Sinclair, B.J. and Sinclair, J., On the use of white mustard-seed for the preservation of health, *The Lancet*, 21 (543), 669, 1834.
48. Masierowska, M.L., Floral nectaries and nectar production in brown mustard (*Brassica juncea*) and white mustard (*Sinapis alba*), (*Brassicaceae*), *Plant Syst. Evol.*, 238 (1–4), 97, 2003.
49. Jackson, L.E., Wyland, L.J., and Stivers, L.J., Winter cover crops to minimize nitrate losses in intensive lettuce production, *J. Agric. Scien.*, 121, 55, 1993.
50. Macdonald, A.J. et al., The use of cover crops in cereal-based cropping systems to control nitrate leaching in SE England, *Plant Soil*, 273 (1–2), 355, 2005.
51. Jensen, E.S., Nitrogen accomulation and residual effects and residual effects of nitrogen catch crops, *Acta Agric. Scand.*, 41 (4), 333, 1991.
52. Martin, H.L., Management of soil borne diseases of beetroot in Australia: A review, *Aust. J. of Exp. Agric.*, 43 (11), 1281, 2003.
53. Dorsainvil, F. et al., Characterisation and modelling of white mustard (*Sinapis alba* L.) emergence under several sowing conditions, *Eur. J. Agron.*, 23 (2), 146, 2005.
54. Balke, D.T. and Diosady, L.L., Rapid aqueous extraction of mucilage from whole white mustard seed, *Food Res. Int.*, 33 (5), 347, 2000.
55. Bernáth, J., *Poppy-The Genus Papaver*, 2nd ed., Bernáth, J., Ed., Harwood Academic Publishers, OPA, Overseas Publishers Association, 1998.
56. Hrishi, N.J., Cytogenetical studies on *Papaver somniferum* L. and *Papaver setigerum* D. C. and their hybrid, Genetica, *S-Grevenhage*, 31 (1–2), 1, 1959.
57. Dobos, G. and Vetter, S., *Variation de Morphingehaltes bei Wintermohn–Herkunften, Zeitschrift für Arznei und Lebensmittelkunde*, 2 (2), 87, 1997.
58. Bavec, F., Oil (seed) poppy, in *Nekatere Zapostavljene in/ali Nove Poljščine (Some of Disregarded and/or New Field Crops)*, Univ. of Maribor, Faculty of Agriculture, Maribor, 2000, 65.
59. Bavec, M., Ekološko kmetijstvo/Organic Farming, (Knjižnica za pospeševanje kmetijstva), Kmečki glas, Ljubljana, 2001, 448 pp.
60. Sharma, J.R. et al., Development of non-narcotic (opiumless and alkaloid-free) opium poppy, *Papaver somniferum*, *Plant Breeding*, 118 (5), 449, 1999.

61. Sharma, J.R. et al., Combating opium-linked global abuses and supplementing the production of edible seed and seed oil: A novel non-narcotic var. Sujata of opium poppy (*Papaver somniferum* L.), *Curr. Sci.*, 77 (12), 1584, 1999.

62. Pelders, M.U. and Ross, J.J.W., Poppy seeds: Differences in morphine and codeine content and variation in inter and intra-individual excretion, *J. For. Sci.*, 41 (2), 209, 1996.

63. Paul, B.D. and Dreka, C., Gas chromatographic/mass spectrometric detection of narcotine, papaverine and thebaine in seeds of *Papaver somniferum*, *Planta Medical*, 62 (6), 544, 1996.

64. Bernáth, J. et al., Variation in alkaloid production in poppy ecotypes: Responses to different environments, *Biochem. Syst. and Ecol.*, 16 (2), 171, 1988.

65. Meadway, C., George, S., and Braithwaite, R., Opiate concentrations following the ingestion of poppy seed products — evidence for the poppy seed defense, *Forensic Sci. Int.*, 96 (1), 29, 1998.

66. Lo, D.S.T. and Chua, T. H., Poppy seed; implications of consumption, *Med. Sci. Law.*, 32 (4), 296, 1992.

67. Moeller, M.R., Hammer, K., and Engel, O., Poppy seed consumption and toxicological analysis of blood and urine samples, *Forensic Sci. Int.*, 143 (2–3), 183, 2004.

68. Selavka, C.M., Poppy seed ingestion as a contributing factor to opiate positive urinalysis results: The pacific perspective, *J. For. Sci.*, 36, 685, 1991.

69. Elsohly, H.N., Elsohly, M.A., and Stanford, D.F., Poppy seed ingestion and opiate urinalysis: A closer look, *J. Anal. Toxicol.*, 14, 308, 1990.

# chapter six

# Alternative fiber, root, and tuber crops

## 6.1 Industrial and edible-seed hemp

### 6.1.1 Introduction

#### 6.1.1.1 History

A historical review of hemp (*Cannabis sativa* L., syn. *C. macrosperma* Stokes, *C. lupulus* Scop.) shows that this recently rediscovered crop was originally cultivated in central Asia, India, and the Nile river valley. It was mentioned as a narcotic in Sanskrit documents at the beginning of 3000 BC. Later, hemp was described as a plant of healing and "joy" in the Sanskrit civilization, mostly as a result of religious celebrations where people became intoxicated from potions made of hemp leaves.

The first records of use in India date back to 2500 BC, and hemp was the first tow plant of the Chinese and Japanese. The Persians, who received hemp from the Punjabi in the upper Indies, used it exclusively as a narcotic. They. Iranians named it "bhanga," a term that meant "being drunk." Babylonians, Egyptians, and Phoenicians were probably not familiar with hemp, because no hemp remains were found in pharaon tombs or in the wrappings of Egyptian mummies.

Hemp gradually spread outward from central and southeastern Asia, especially as people migrated westward. Hemp was used by Indo-Persian tribes, who received it from old Arabia and the Caspian countries of Bactria and Sogdiana. Thracians produced hemp for tow and and passed the practice on to the Greek and Roman civilizations; hemp was also produced in Gaul. In the second part of Homer's *Odyssey*, written in the sixth century BC, boat equipment consisting of ropes and sails is mentioned. In part four, Jelena, wife of the emperor Telemachus, poured him and his companions "a miraculous juice sweetened with tranquilizers to help them forget their sorrow." Historians believe that these passages refer to hemp. Povolžje, in eastern

*Figure 6.1* Hemp brewage.

Europe, was another important center of hemp production; there it was produced by Hungarian and Finnish tribes, and later by Bulgarians [1].

In records on the Spartans, King Hieron II. Syracose was said to have used hemp fibers and wood resin on his quest (287–212 BC). Herod's records from the fifth century BC state that Skits, the presumed ancestors of the Slavs, produced hemp that was similar to flax but much thicker and taller. Hemp is sown there but is also grows on its own. Thracians made canvas out of it that could not be distinguished from flex canvas. Jevtić [1] writes that narcotics were produced from hemp as far back as centuries BC. Botanists and historians believe that the Skits played a very important role in spreading hemp into Europe and Little Asia. In the first century BC, hemp used for fibers was mentioned by the Roman writers Lucius Julius Kolumela and Markus Terencius Varon; Plinius also mentioned hemp as a healing plant in the book *History of Nature*. He wrote that in fertile soil, hemp could reach the height of trees.

Between the second half of the first century to the early Middle Ages, hemp production experienced no noticeable developments. There are notes from the period of Karl the Great, however, mentioning hemp as a very useful plant. Similarly, the Russian Duke Oleg equipped a fleet of 200 ships on his march to Constantinople, an achievement that probably have been impossible without toughly sails made from hemp fibers.

By the beginning of the Middle Ages, hemp was commonly as a narcotic in the area of Arabian Caliphate, i.e., in northern Africa and part of Asia. Hemp's widespread use was often contributed to by the fact that the Koran prohibits drinking wine; the most available drug was hashish, which was often abused. It was not by chance that one of the Arabic tribes was known as the "Hashishians." The word "hashish" is an old Iranian word for "grass." According to other sources, an eighth-century Muslim sect of fanatical killers in Arabia were called "hasiasina," a name derived from the French and English words for "assassin."

When the Crusades ended, trade ceased between Russia and Tatar-Mongolia, which resulted in a reduction of hemp production for fibers. Around the same time, production began to spread in the western part of Europe. Cities like Venice and Genova reached the peak of their development, army fleets also strength, and hemp production quickly spread through Italy; Italy became a great hemp exporter. Growing economic power in Spain, England, and Holland allowed hemp production to spread to other European areas as well. The pressure of exporting competition began to cause problems, though, leading Pope Sikst V. to issue a written demand in 1856 setting quality standards for hemp fiber intended for export. As northeastern Russia recovered from the sixteenth-century Tatar-Mongolian, it once again began to produce and trade larger amounts of hemp. By the beginning of the seventeenth century, they produced enough for their needs and exported it to England and other countries [1*, 2].

The growth of hemp was first noted in America in the sixteenth century. The produced low-growing Scottish tall-type hemp in favorable growing conditions; this type of hemp differed from the southern types [3].

France was the most important hemp producer by the end of the eighteenth century, and by the early nineteenth century, it was being grown on 100,000 hectares of land. The introduction of steamboats and a reduction of hemp prices in Russia caused hemp production to drop overall in Europe. It was sown only in Italy and Hungary, and production fell across the continent. In ex-Yugoslavia, hemp was grown on 100,000 hectares of land in 1949; that number fell to less than 5000 hectares in 1991 and only 1000 hectares in 1998.

Throughout its history, hemp growth spread with nomadic tribes and nations. It served many purposes; in northern parts of Europe, it was used as a fiber plant, an oil plant, and a grain plant grown for nutrition. In Asia (China, northern India, Pakistan, Afghanistan, and Turkey), parts of Africa, and North and South America from the sixteenth century onward, hemp was produced mainly as a hallucinogenic drug and only rarely as medicine.

* Reprinted from Hemp – *Cannabis sativa* L, in *Posebno ratarstvo 2*, Jevtić, S., Jevtić, S., Ed., 200, Copyright (2006), with permission from Naučna knjiga.

Oil lobbies achieved the prohibition of hemp production in developed regions of America and Canada, due to propaganda that industrial hemp was associated with the use of illegal narcotics. In Europe, the female flowers *Cannabis sativa* L. var. *indica* were used for drug production only in the eighteenth century. The widepread production of this drug led the United Nations Organization (UNO) to hold a convention in 1961 for the purpose of fighting drug abuse. Consequently, depending on countries' varying laws, only cultivars with 0.20 to 0.35% THC (delta-9-tetrahidrokanabinol) may be produced. Since cultivation of fiber hemp has been permitted in some countries during recent decades, a number of surveys were published concentrating on the suitability of this plant for cultivation and industrial processing, and the interest in hemp byproducts is increasing. Hemp has been rediscovered as an interesting crop with a large plasticity, which allows it to be grown under a wide variety of agroecological conditions.

## 6.1.2  Botany and ecotypes

Hemp is an annual plant. It belongs to the family *Cannabaceae* and the genus *Cannabis*, with only one species of common (industrial) hemp *Cannabis sativa* L. Many botanists have conducted experiments in order to try and describe three species based on this monotype genus and polymorph species. Hemp (*Cannabis sativa*) is divided into two key types: European and East Asian. The European type is further divided into north Russian, central Russian, and southern types; the East Asian type is divided into Chinese, Japanese, and coastal subtypes. North Russian hemp is located in the extreme north and is therefore the earliest and least-branched type. Male plants ripen in 30 to 35 days, and female plants ripen in 60 to 80 days. Plants reach a height of 0.5 to 0.6 m and have small seeds with 12 to 15 g of 1000-seed weight.

Central Russian ecotypes are medium-early with a growing stage lasting between 110 and 115 days. Thin and branched stems reach a height between 1.3 and 2.0 m. Large leaves contain five to seven small leaves, and the 1000-seed weight is between 15 and 20 g. This type of hemp is very common.

The southern ecotype is also known as the Italian ecotype and ripens in 125 to 140 days. Plants grow to between 2.5 and 4.5 m high, and the 1000-seed weight is 18 to 26 g. Compared to other types, the southern type is considered more fertile due to the longer growing stage.

East Asian hemp has the longest growing stage, which lasts more than 140 days, and the plants reach a height of 4.0 to 6.0 m.

Differences among ecotypes are not so clear anymore, due to the presence of certain genetic structures in cultivars of individual breeders.

## 6.1.3  Plant morphology and anatomy of stalk

Growth and developmental morphological differences in hemp are caused by sexual dimorphism between male and female plants. Some new genotypes are also mutual.

Roots are well developed. The root system consists of a taproot with branched roots of second and third types. Taproot reaches a depth of 2.0 m, while lateral roots reach a depth of 0.6 m. The root system of female plants is usually more developed than the root system of male plants. The highest root mass can be found at a depth of 0.3 to 0.4 m, and the roots grow very intensively at the fast growth stage. The root weight is considerably lower than the weight of the aboveground part; this is the reason for lower yield when nutrient deficiencies or moisture exist. Hemp was formerly used as a test plant for uniformity of soil on test plots.

### 6.1.3.1 Stem

Plants can reach a height of between 0.5 and 5.0 m, depending on the basic ecotype and methods of crossbreeding; additionally, male plants grow higher than female plants. Stem diameter varies from 2 to 40 mm. Stem height and diameter depend on the ecotype, climate, fertilization, size of growing space, and other cultivation circumstances. The stem represents between 60 and 65% of the aboveground weight, and its length can increase 5 to 7 cm daily. Young stems are soft, juicy, and hairy, while old stems are firm and wooden. The stems contain fibers, where hemp is mostly grown. In mature plants, a layer of fiber sheaves is easily separated from the wooden part. Stem diameter can also vary on the same stem, according to the height at which it is measured. Near the ground, the stem is round, and the central part is hexagonal. In the middle of the stem height, furrows are visible that deepen toward the top. The stem is hollow throughout the length, except for the extreme ends, and it sometimes branches on the top (except in cases of high-density sowing). Variations in plant population may result in different branching and stem length, causing a variable quantity and quality of fibers. Stems consist of 5 to 20 internodes, and the internodes of male plants are longer than internodes of female plants. An increase in the number of internodes results in higher fiber quality.

The anatomy of stems (from the outer to the inner part) is as follows:

1. Outer stem layers consist of cover tissue. Cover tissue protects the plants from temperature shocks, mechanical damage, and excess transpiration, and plays an important role in gas exchange. Tissue consists of epidermis with one layer of tightly fitted cells, which are covered by cuticle. Constituents of cuticle are waterproof wax-like substances — cutin.

2. Primary crust consists of three tissues. Immediately below the cover tissue lies mechanical or supporting tissue called collenchymas, built of cells normally developed in all directions. At the edges of stem the collenchymas cells are joined in compact sheaves. These cells provide firmness to the stem, which is very important in the first development stages until the formation of fiber and lignifying of inner cells. Under the collenchymas is parenchyma, with thin cell wall and intercellular spacing. It has a diverse role of assimilating,

conducting, stocking, and so forth. Under the layer of parenchyma cells is starch endoderm, with one layer of starch grains. These reserve substances enable the plant to survive before the green plant part is ready to assimilate.

3.   Primary fibers serve as a basis for yarn fibers. They consist of thin cellulose cell walls (pericycle parenchyma) and cells with thickened cell walls. Fibers are extremely elongated (prosenchymatous), which gives them firmness and more elasticity than wooden fibers. Primary fibers often join in sheaves. Parts of their cell walls can turn wooden with plant maturity.

4.   Under primary fibers are phloem fibers and secondary fibers. A composition of sieve tubes, secondary fibers, and parenchyma cells exist. Sieve tubes of hemp can be depicted as capillaries with perforated barriers and cell walls that are neither thickened nor lignified. Sieve tubes are used for transporting organic substances in plants. Secondary fibers are similar to primary ones but their cell walls are thinner, less developed, and often appear in disorder. Sometimes they are located in compact or denser distributed concentric layers or sheaves. The secondary fibers are less firm than the primary ones. Primary and secondary fibers constitute rough wooden cells near soil and, in production, the bottom 5 to 10 cm are thrown away. The ratio of primary and secondary fibers is conditioned by variety and the technology of production. With high plant density, there are fewer secondary fibers, whereas with low plant density, the amount of secondary fibers increases up to four times. The smaller secondary fiber layer is a consequence of a delayed sowing date. Mediavilla et al. [4] and Schäfer and Honermeier [5] have shown that the formation of secondary fibers starts at the beginning of the flowering stage and thus fewer secondary fibers are formed, at a lower degree of ripeness.

5.   Under schlerenchyma a thin layer of tender cells capable of division exist, called cambium. Toward the center of the stem, new wooden cells are formed; in the opposite direction, secondary fibers are formed. Tender cambium cells are a barrier for easy separation of stem bark and stem core. The fiber quality in the bark (mainly primary bast fibers and some secondary bast fibers) is much better than in the core (mainly high-lignin, libriform fibers). The appropriate bark:core ratio can be realized by aiming at high plant populations. The bast fiber content of the stem increases with plant population.

6.   Under cambium there is a core part of cells and water vessels (xylem). Xylem consists of long capillaries with thickened and core cell walls that transport nutrients from soil. Core fibers that support the plant are mostly dead cells.

7.   The inner stem part is often hollow and cannot be called pith in the real sense of the word. It consists of parenchyma cells, which can be dead or capable of division with young plants.

The morphology of the fiber cells of hemp can be significantly affected by weather conditions, location, and seed density. These same properties probably also affects the mechanical characteristics of hemp fibers [6].

Hemp leaves consists of petiole and blade. The first two leaves are simple and poorly developed, with jagged edges at the green cotyledons. The second pair of true leaves are triple-palmately lobed. The leaves that follow are grown from 4 to 9, or even 13, small parted leaflets. Central Russian types have 5 to 7 leaflets, while southern types have 9 to 11, or sometimes as many as 13, leaflets. The number of parted leaflets decreases toward the top of the plant. On the top, small to medium simple pointed and jagged leaves form. Leaves are hairy on both sides, and they are placed alternately only at the top part of the stem. The leaves' color can be intensely green to light green.

Flowers are self-pollinated and plants are dioecious, though monoecious varieties are gaining value throughout the world. Hemp is almost exclusively wind-pollinated. The flowering of female plants begins 2 to 3 days before the flowering of male plants; according to the ecotype, this difference can extend up to 10 days. Female plants flower after 15 to 30 days. The stage from fertilizations to full seed maturity lasts 40 days. From the opening of the first flowers to the flowering of all plants requires 5 to 10 days. The flowering stage produces a lot of pollen, with a couple hundred flowers opening daily on one plant. In cold weather, flowers remain closed, which causes poorer fertilization. Female flowers flower in the morning and are pollinated rather fast. Male plants (sometimes called white plants) mature prior to female plants (named black plants) following pollen formation.

The male flower consists of the pedicel, flower carpel, and stamen. Flower carpel consists of five elongated, inward-turning leaflets of pale green color. There are five stamens with long stamen filaments, so that anthers hang out of the flower.

The pistil of the female flower with superior ovary contains one seed, and the pistil has two parallel stigmas. It darkens 1.5 to 2 days after fertilization. Every female flower is surrounded by fine, hairy bracteole. The THC share with *Cannabis sativa* var. *indica* increases to the flowering stage and can be found in the highest quantities in bracts of female flowers.

With monoecious plants male inflorescence is placed together inflorescence on the top of the stem, and male flowers are joined in spikelet inflorescence in the nodes of leaves.

Fruit (we usually talk about seed regarding yield and harvest) is a two-part achene that is wide with a round shape. The calyx is quite hard and protects the seed mechanically. The seed is brownish-green, grayish, or silver grey to black in color, with more or less expressed mosaic spots. The 1000-seed weight varies according to ecotype from 12 to 20 g. The seed contains 30% of fat and 20% of protein.

*Table 6.1* Definitions and Codes of Growth Stages of *Cannabis sativa* L. Plants According to Mediavilla et al. [26]

| Code | Definition | Remarks |
|------|-----------|---------|
| | **Vegetative Stage** | |
| 1002 | First leaf pair | |
| 1004 | Second leaf pair | |
| 10nn | nn/second leaf pair | |
| | **Flowering and Seed Formation** | |
| 200 | GV point | Change of phyllotaxis on the main stem from opposite to alternate. |
| | **Male** | |
| 2100 | Flower formation | First closed staminate flowers. |
| 2101 | Beginning of flowering | First opened staminate flowers. |
| 2103 | End of flowering | 95% of staminate flowers open or withered. |
| | **Female** | |
| 2200 | Flower formation | First pistillate flowers. Bracts with no styles. |
| 2201 | Beginning of flowering | Styles on first female flowers. |
| 2202 | Flowering | 50% of bracts formed. |
| 2203 | Beginning of seed maturity | First seeds hard. |
| 2204 | Seed maturity | 50% of seeds hard. |
| | **Senescence** | |
| 3001 | Leaf desiccation | Leaves dry. |
| 3002 | Stem desiccation | Leaves dropped. |

*Source*: Reprinted from Mediavilla, V. et al., *J. of Int. Hemp Assoc.*, 5 (2), 67. Copyright (2006), with permission from International Hemp Association.

## 6.1.4   Ecology of growth and development

Growth and development of hemp can be roughly divided into the following stages: germination, seedling, slow growth, intensive growth (appearance of first flower buds), flowers and flowering, fruits, and maturity. New definitions and codes of growth stages are described in Table 6.1.

There are 9 stages characteristic of the organogenesis of generative organs for male plants and 12 stages for female plants.

The highest aboveground weight is created between the bud development stage and the end of the flowering of male plants. The male plants tend to flower and senesce earlier than the female plants. This means that growing conditions need to be highly suitable for quite a short period; this depends on plant ecotype or variety characteristics, however, because plants with a longer growing period create more aboveground weight. With early cultivars, the grain yield is higher. Plants with a longer growing stage are taller than plants with a short growing stage. When introducing new cultivars in different growing conditions, the amount of stem and fiber production often decreases, whereas seed yield increases.

The growing stage for breeding plants originating from eastern countries lasts between 140 and 160 days, andfort the temperature sum must be 3500°C. Northern ecotypes have growing stages from 60 to 90 days, and their temperature sum demand is from 800 to 1000°C. The earliness or lateness of French cultivars is expressed in numbers; for example, Ferimon 12, Fedora 19, Felina 34, Fedrina 74, Fibrimon 24, Fibrimon 56, and so forth. Higher numbers indicate that the cultivar is late. The growing period length is tightly correlated with the length of individual development stages. This is not true, however, for differences among ecotypes and cultivars from the emergence of seedlings to the appearance of three pairs of leaves. After the appearance of three pairs of leaves, male plants grow faster and develop more plant dry matter within a certain time unit. Intensive growth begins about 40 days after emergence, while additional growth of dry matter and fibers take place in the sixth decade. The length of the stage depends on the length of the growing period of the individual cultivar. Growth almost stops after flowering in male plants. With female plants, however, the growth of vegetative parts can be reduced due to poor nutrient provision at grain filling. In this case, their own spare resources are translocated into the seed.

Growth and development of hemp are conditioned by nutrient and moisture availability in soil, especially from weeks 4 to 10, when additional growing of dry plant matter is the highest. Hemp demands suitable climate and soil conditions. Infertile, sandy, cold, or too moist soil is inappropriate. Temperatures also greatly influence hemp growth. In 20°C, hemp can grow 4 cm day$^{-1}$ but in 10.2°C, hemp can only grow 0.5 cm. In very favorable growing conditions (of moisture, temperature, and nutrients), hemp has been noted to grow as much as 12 cm per day$^{-1}$.

The minimal germination temperature is 4 to 5°C, although hemp will germinate between 1 and 2°C. The optimum laboratory germination temperature is 20°C. Changing temperature accelerates germination; young seedlings are not harmed by a sudden drop of temperature to –15°C.

Longer periods with subzero temperatures can destroy germinated seeds, especially with more than 80% field soil capacity for moisture. Emerged plants react differently to changes. Early sowing or cold climate areas can present an obstacle for germination. They can survive low periods of frost from –1 to –5°C. The crop already forms full ground cover after a thermal time of about 400 to 450°C. Spring temperatures can influence gender formation. Low temperatures from –5 or –6°C accelerate formation of monoecious plants with otherwise diecious plants; the number of monoecious plants in such populations can vary from 1 to 50%. At flower bud stage, we can differentiate gender of plants that are destroyed by subzero temperatures.

Sufficient moisture is the most important part of successful hemp growth in comparison to other crops; the shortage or excess of water during early growth stages can be a destructive factor for successful production.

Hemp behavior varies in different eco-regions, and the plants may not be able to make full use of the potential of the seasons all over the world. Ecotypes originating from eastern countries are long-day plants. They flower quite late and form only a few flowers. Northern ecotypes react to short days and consequently have lower dry matter formation. The cultivars grown in Europe have a critical photoperiod between 14 and 15.5 h.

## 6.1.5   Cultivation practice

The main problem facing hemp cultivation might be crop establishment: hemp is very sensitive to a lack of available nutrients from the soil, unequal or poor soil structure, and a shortage or excess of water during early growth stages.

### 6.1.5.1   Cultivars

Key cultivars in the EU assortment are as follows: Carmagnola, Delta Liosa, CS, Delta 405, Fedora 19, Fedrina 74, Felina 34, Ferimon, Fibranova, Fibrimon 24, Fibrimon 56, and Futura — standard 4/96. The cultivars Bialobreskie, Kompolti, USO 11, USO 13, YUSO 14, and YUSO 1 have less than 0.3% THC.

Under extremely dry and warm growth conditions, some cultivars — such as Secueni 1, Unico B, Kompolti Hybrid TC, and Beniko [8] — might contain up to 0.8% delta-9-tetrahydrocannabinol. In Australian growth conditions, the cultivars Kompolti, Unico B, and Futura 77 had a content of delta-9-tetrahydrocannabinol below the legal maximum of 0.35% (dry weight basis) [9].

Cultivars may be separated according to target production. Slovenian conditions for fiber (t), seed yield, and oil production (y) propose the following cultivars: Kompolti hibrid TC (y), Kompolti kender (t), Kompolti sazru (t, y), Novosadska konoplja (t), and Unico B (y).

Hemp has almost no limit demands with respect to crop rotation. It is usually sown after cereals, annual legumes, other arable field crops, and grass or cabbage. It is one of the plants that can often return to the same field; when fertilizing with stable manure, it can be grown as a monoculture. Hemp is a suitable precrop for cereals or can be sown after true cereals due to timely classical soil preparation. By introducing hemp into crop rotation, we can successfully suppress weeds in cereal crops, especially if crops are produced that allow for intensive reproduction and growth of weeds.

Results show that leaves contain 4.6 times more nitrogen, 3.2 times more phosphorus, and 1.7 times more potassium than stems. The average nutrient content in stems differs according to the stem part of hemp (Table 6.2).

The highest hemp demand for nutrients is noted in the first half of the growing stage. Root system weight at this stage is much lower than the aboveground part. Consequently, roots are not developed enough for nutrient uptake, which enables intensive growth and the available necessary nutrient amount in poor soil. In 100 kg of aboveground dry matter weight

*Table 6.2* Average Nutrient Content in Individual Parts of Mature Hemp Stem [2, 4, 5]

| Plant Part | % in Dry Matter | | | | | |
|---|---|---|---|---|---|---|
| | N | P | K | Mg | Ca | S |
| Leaf | 2.40 | 0.42 | 1.77 | 0.59 | 0.81 | 0.45 |
| Bark | 0.57 | 0.22 | 1.06 | 0.30 | 0.32 | 0.35 |
| Stem | 0.52 | 0.13 | 1.06 | 0.12 | 0.32 | 0.36 |
| Together | 1.13 | 0.22 | 1.17 | 0.72 | 0.47 | 0.39 |

of hemp, various sources show the following nutrient amounts in the soil: 1.0–2.9 kg N, 0.22–0.75 kg $P_2O_5$, and 0.83–2.74 kg $K_2O$ [4].

If we speculate on the basis of a published paper, the available nitrogen should be a minimum of 225 kg N ha$^{-1}$ for 22,500 kg of aboveground matter ha$^{-1}$ yield. Once analyzed, available nitrogen may exist solely for orientation purposes; however, in those cases, adequate input of organic nitrogen and potential mineralization are necessary.

Recommendations for seed production without soil analysis in Austria are the following: 100 kg N ha$^{-1}$, 80 kg $P_2O_5$ ha$^{-1}$, and 140 kg $K_2O$ ha$^{-1}$ for fiber production, the recommendations are: 80 kg N ha$^{-1}$, 80 kg phosphorus ha$^{-1}$, and 120 kg potassium ha$^{-1}$. In European regions, recommended nutrient amounts vary according to composition of soil and growing conditions, from 45 to 150 kg N ha$^{-1}$, 45 to 110 kg $P_2O_5$ ha$^{-1}$, and 45 to 110 kg $K_2O$ ha$^{-1}$. The efficiency of the nitrogen fertilization rate depends also on available mineralized nitrogen, for the useful nitrogen rate depends on potential mineralization. If fertilization occurs with slurry or liquid manure, these fertilizers are clogged in the soil before sowing. Manure or compost can be plowed in autumn. In heavy or poor soil, we manure with phosphorus and potassium manures after basic soil cultivation in autumn, and part of them are used before or during sowing. With nitrogen, we manure before sowing or at the stage of three pairs of leaves, at the latest. Later additional fertilization — especially with high nitrogen amounts — negatively influences fiber quality or the height of plants, which can be too high for seed combining.

Autumn plowing is very suitable for hemp. Late spring sowing should cause the emergence of weeds, which are mechanically destroyed by presowing cultivation. Early spring is the appropriate time for sowing; later sowing is not recommended. Nevertheless, on the basis of experiments in Austria, the highest seed yield was achieved with sowing dates to the middle of springtime; later sowing did not result in maturing, even with early cultivars [12]. Sowing date trials with nine hemp cultivars were conducted across two seasons, incorporating dates between early and late spring sowing and a single autumn planting. Stem and bark yield declined with delays in sowing after the middle of spring in response to a decline in calendar days and thermal time from sowing to flowering. The response was most pronounced in sowings of Kompolti, which flowered within a short period and differed more substantially in durations to flowering [9].

Sowing depth is 2 to 4 cm or more in humus light soil. The energy of germination and germination force are described quite differently in literature on the subject. This depends on the 1000-seed weight, seed germination, and soil characteristics. In organic production, 8 to 15 kg seed ha$^{-1}$ are recommended to be sown at 0.4 m interrow spacing, which enables interrow hoeing or cultivation. With low cultivars intended for oil plant production, the sowing of 13 kg seed ha$^{-1}$ is recommended. In classical production, 12 to 20 kg seed ha$^{-1}$ should be sown with interrow spacing of 22 cm in every other row. With high plants, the spacing should be between 0.5 and 0.7 m, and 12 kg seed ha$^{-1}$ should be used. In such cases, it is best to cultivate every second or third line with double rows in-between. With short cultivars, 40 kg seed ha$^{-1}$ should be sown.

In general, high plant populations without missing plants in the plant stands may give the best fiber quality. In high plant populations, the large plants suppress smaller ones, and this may even result in self-thinning. The influence of plant population on hemp yields was investigated in a few cases. In Australia, the investigated densities were from 50 to 300 plants m$^{-2}$. Plant density was most pronounced for populations of 200 and 300 plants m$^{-2}$, but the final harvest stem yield responded in a parabolic manner to plant density, with maximum yields at about 110 plants m$^{-2}$ [9]. The stem yields and quality depend on inter-plant competition influenced by dense plant population [13], especially at 270 plants m$^2$, and high nitrogen rate [14] causes self-thinning. Depending on growth conditions, differences in the percentage of the long, high quality bark fiber at final harvest were generally small and not significant [9].

Weed management is not a problem with hemp, because hemp suppresses weeds efficiently as one of the most weed-competitive plants. Hemp production for oil with longer interrow spacing enables the germination and development of some absolute weeds (*Cirsium arvense* L., *Sorghum halepense* L.) and some relative weeds.

## 6.1.6  Harvesting, storage, and processing

The official suggested harvest time for fibers is 2 to 3 weeks after flowering. Based on the results of the physical properties of the bark, Keller et al. [15] also concluded that the optimal harvest time is at the beginning of seed maturity, which is approximately 3 to 4 weeks after male flowering; in spite of that, the maximum yield of the stem, bark, and fiber was reached at male flowering ("technical maturity") [4].

Special machines are used for harvesting, and these machines should be bought by producers joined together in an association. Dry stems or barks with lengths of 0.5 to 0.7 m can be cut by using an adjusted machine otherwise used for flax. Dry plant material is packed, according to contractor demands, in bales or bundles.

The average yield of stem dry matter in agriculture practice varies between 2000 and 5000 kg ha$^{-1}$, sometimes reaching even higher. In field micro trials, the yield can reach 12,000 kg ha$^{-1}$. Higher yield in trials is almost always a consequence of experimental errors due to a lack of consideration of edge influence or too-small plots [11; Starevi, personal communication and author experience]. But according to Struik et al. [16], conventional grown fiber hemp may yield up to 22,500 kg aboveground dry matter per hectare (20,000 kg stem dry matter ha$^1$), which may contain as much as 12,000 kg ha$^{-1}$ of cellulose, depending on environmental conditions and agronomy. Some of the same cultivars used in previous experiments (Kompolti, Unico B, and Futura 77) had the highest single-plot dry stem yields in Australian trials (up to 15,000 kg ha$^{-1}$ and bark proportions up to 40%) [8].

Hemp harvesting for bird food and oil is executed at the stage of full maturity when the seed can be shaken from the top of the plant. Combines may be used for harvesting.

Machines need to be adjusted for high cultivars. For grain, only the top of the plant can be harvested; the rest can be harvested for fiber. Seed yield in Austria in 1995 and 1996 amounted to between 800 and 1300 kg ha$^{-1}$, while yields in Belgium and France in some cases exceeded 2700 kg seed ha$^{-1}$.

Hemp seed is dried after harvesting to 10% of dry matter, and the drying temperature must be less than 50°C. Seed used for the cold-pressing of edible oil should not contain any crop protection substances and fungi contamination. It is stored and sold in paper or jute bags.

## 6.1.7   Utilization: special organic products

Hemp is valued for its rough, firm, and water-resistant fibers; it is used in the rope industry to make fish nets, different fabrics, and tents. Fibers are also useful in the paper industry, as a construction material, as fuel, and more. The core part is used for making artificial fibers, cartons, and construction material. During recent decades, a very extensive description of hemp utilization for industry processing and marketing was done, providing for its use in textiles and geotextiles [17–19], high-quality papers [20, 21], composites for the automotive industry [22, 23], and as insulation materials for houses [24, 25]. The versatility of the seed lends itself to the development of numerous products for the food, cosmetic, therapeutic [26], beverage, and nutraceutical industries. The quality of oil is currently under investigation in order to improve the economic and/or environmental performance of an unconventional crop through innovative uses of its components and byproducts. As a result of the processing of hemp fibers, the seed becomes an interesting byproduct [27].

Organically produced and processed products deserve special attention. Hemp may be utilized for biotextile insulation and as building blocks for biohouses; as an energy crop for heating; and as edible seeds and oil, in beverages and cosmetics, and so forth. The highly polyunsaturated oil of hemp seed has been used for printer's ink, wood preservatives, and detergents and soaps [28].

In the case of using bast fibers of flax and hemp for insulation, it is essential to follow good manufacturing practices and to keep the insulations in a dry place throughout the manufacturing and building process in order to minimize hygienic risks (fungi and bacteria) in insulations [29].

Technical fibers are of different quality. The fiber of the male flower plant is soft and strong and can be compared to flax fiber for making canvas. Female plants have rougher fiber, suitable for making ropes. The fiber morphology and chemistry are influenced by the growth stage, the age of the plant, and fiber processing. The fibers of green hemp are soft and thin, with mechanical properties similar to flax fibers. Lengths of primary and secondary fibers also differ. Primary fibers are 8 to 40 mm long, while secondary fibers are no longer than 4 mm. Primary fibers are joined in sheaves, which represent technical fibers after splitting from other tissues. The lengths of technical fibers vary from 0.2 to 0.5 m or more, according to stem length. The highest yield from technical fibers comes from the middle and top plant part. Yield can be between 15 and 30%, though 20% is difficult to exceed (male plants are the exception).

The amount and quality of technical fibers are conditioned by stem quality: longer stems result in longer fibers and better yield. Thinner branch axles are to be expected among higher plant populations. This is an advantage in fiber production because hemp fibers are phloem fibers, which are located directly beneath the epidermis. In thinner stems, they can be separated more easily and are much more suitable for the textile industry [30]. This condition may also be reached by dense plant populations, which cause strong elongation of the primary bast, producing long low-lignin fibers. Branched stems negatively influence fiber quality and yield; yield is lower with thick stems.

Firmness and color also depend on time of harvest and method of after-treatment. A promising method for obtaining fine hemp fibers is the controlled biological degumming of decorticated bark in bioreactors using adapted microorganisms and their enzymes [31]. Kelle and Leupin [15] considered the mechanical decortication of green dry stems without degumming of the bark; the results revealed that a harvest time at the beginning of seed maturity leads to easier decortication without any effect on the tensile strength of the bast. For decortication of fresh stems, including a subsequent degumming process, a harvest after the flowering of the male plants results in fiber losses during decortication and fibers of reduced fineness. Hemp fibers may be dirty yellow, green-grey, metal grey, or almost white in color.

In order to revive the use of hemp in the textile industry, new technical and technological solutions are necessary [32]. Special attention has been given lately to the growth of hemp, which is much quicker than forest increment. While pine forest increments produce 2.5 $m^3$ $ha^{-1}$, 10 to 12 $m^3$ of wood from a hecare of hemp are produced in comparison. Around 65% of hemp stem weight consists of wood, which is constituted of 40 to 48% cellulose, 26% lignin, and 32% pentosans. Its caloric value is no less than 15.74 MJ $kg^{-1}$, which is more than is found in wood (11.30 MJ $kg^{-1}$) and less than coal (20.09 MJ kg$^{-1}$). The wooden part of hemp has good thermoinsulation properties in construction and potential in the chemical industry for furfural synthesis. Hemp has been receiving a lot of attention as a possible fuel and construction material from environmentalists and organic producers. Ashes from wooden hemp parts should also be mentioned, due to the content of 24% CaO, 4,85% P$_2$O$_5$, and 6,3% K$_2$O.

### 6.1.7.1   Hemp seed

As reported by Callaway [33], hemp seeds have not been studied extensively for their nutritional potential in recent years, nor has hempseed been utilized to any great extent by the industrial processes and food markets that have developed during the twentieth century. Hemp seed contains 20 to 30% carbohydrates, 25 to 35% oil, 10 to 15% insoluble fiber, and a rich array of minerals [33–35] and vitamin E [36]. Hemp seed also contains 20 to 25% proteins, mainly edestin and albumin. Both of these high-quality storage proteins are easily digested and contain nutritionally significant amounts of all essential amino acids. In addition, hemp seed has exceptionally high levels of the amino acid arginine [33].

Hemp seed can be used for poultry feed and is also appreciated as food for birds. After oil pressing, 65% of the oil seed cake weight remains from seed. Approximatelt 1 kg of oil seed cake can replace 4.5 kg of corn silage from whole plants. Oil seed cakes are high-quality protein feed because 1 kg contains as many digestible proteins as 2.85 kg of oats, 3.0 kg of barley, 3.2 kg of corn grain, or even 25.3 kg of corn silage from whole plants.

Hemp seed can be used for making "hemp" bread in a mixture made with 15% hemp share. Besides seed and fiber production, hemp can also be produced for fuel. For these purposes, only cultivars without THC content or with the allowed THC content are produced. For production and marketing, only cultivars with less than 0.3% THC may be used.

Hemp seed, in addition to its nutritional value, has been used to treat various disorders for thousands of years in traditional oriental medicine. It has demonstrated positive health benefits, including the lowering of cholesterol and high blood pressure [37]. Recent clinical trials have identified hemp seed oil as a functional food, and animal feeding studies demonstrate the long-standing utility of hemp seed as an important food resource [33].

The oil, because of this feature and the presence of linolenic acid, is ideal as an ingredient for light body oils and lipid-enriched creams, known for their high penetration into the skin [28].

Hemp oil's iodine value is from 140 to 167, with an average of 159. With mechanical cold-oil pressing, we get 28% of oil. Refined hemp oil is ranked among quality oils on the basis of its color and taste. Most often, it is used for the preserving of fish and in flour confectionery industries. As an easy-drying oil with unsaturated fatty acids, it can be used in the production of soap, varnish, and oil colors. Oil can be used for nutrition, and the production of edible hemp oil is unlimited in Europe. Cold-pressed oil tastes like grass juice, making it suitable for culinary purposes. The price of hemp oil is somewhere between olive and sesame oil.

Hemp seed oil has been deemed perfectly balanced for human nutrition in regard to being a rich source of two essential polyunsaturated fatty acids: linoleic acid (18:2 omega-6) and alpha-linolenic acid (18:3 omega-3). The omega-6 to omega-3 ratio (n6/n3) in hemp seed oil is normally between 2:1 and 3:1, which is considered to be optimal for human health [33]. In addition, the biological metabolites of the two essential fatty acids, gamma-linolenic acid (18:3 omega-6; GLA) and stearic acid (18:4 omega-3; SDA), are also present in hemp seed oil.

The industrial processing of edible oil has also begun. An increase in the use of hemp oil for salad dressings has been noted, and microwave heating has recently been found to be particularly effective in releasing membrane-bound tocotrienol and tocotrienol-like compounds and in maximizing the stabilization of biological materials against degrading enzymes [38]. Microwave treatment of hemp seed produces positive beneficial changes in the quality of hemp oil. The elevated tocopherol concentration suggests an improvement in hemp seed oil quality that can become an economically and environmentally sound resource for the functional food and nutraceutical industries [27].

## 6.2 Flax

### 6.2.1 Introduction

Flax (*Linum usitatissimum* L.) is one of the oldest cultivated plants known to mankind. It was grown in Neolithic times in Switzerland and southern Germany, and in Spain during the Bronze Age. Egyptians grew it in 3000–4000 BC and were followed by the Romans, Gauls, and Celts, who were the beginners of production in western Europe; Slavic nations took over production from the Greeks.

The ancestor of flax was pale flax (*Linum angustifolium* L.). The primary center of fiber flax is southeastern Asia and for oil (seed) flax is northern Africa. A lot of fiber flax is produced in Northern Europe at the Atlantic coast (The Netherlands, England, Belgium), at the North Sea (Denmark, Sweden), and in the Baltic States. Moderately cold and moist climates suit flax well. Oil (seed) flax grows better in the dry and warm climates of Southern Europe, Africa, India, Argentina, and so forth, though Russia produces 65 to 70% of yield worldwide. The largest importer of flax grain is Great Britain [39].

Flax production has emerged as a potentially interesting undertaking in a wide range of agroecological environments, also as an efficient plant for remediation heavy metals from the soil. There are several advantages that could result from flax being produced organically, as well for textile and edible seeds.

## 6.2.2   Botany and cultivars

Flax belongs to the family of *Linaceae*. Of 200 species, the most important for production is the Eurasian subspecies of ordinary (cultured) flax *Linum usitatissimum* L. According to the branching, stem height, and so forth, flax is divided into fiber flax (*Linum elongatum*), oil (seed) flax (*Linum brevimulticaulis*), and intermediate flax (*Linum intermedium*) [39].

Flax varieties suitable for Slovenian conditions are not officially verified, in most cases. Cultivars of fiber flax are Belinka, Natasja, and Regina. From the Netherlands come the very-well-known cultivars Wiera (the oldest one), Diana, Noblesse, and Solido, which are suitable for spring sowing. The French early cultivars of oil flax are Antares, Mikael, Ocean, Linda, Atlante, and the winter cultivar Nivale. The German early cultivars are Ceres, Hella, Kreola, and Liflora. There are also Ariane, Barbara, Blue Cip, Hungarian Gold, Ran, Kiszombori 41, Linneta, Gregor, Norlin, Omega, Sandra szegedi 41, and Sandra szegedi 62.

Lisson and Mendham [40] reported that European flax cultivars yielded significantly more stem and bark fiber than the Australian flax cultivars. Of the former group, Ariane (841 g m$^{-2}$) and Marina (883 g m$^{-2}$) performed the best in terms of stem yield production, while Viking had comparable bark yields to these two cultivars.

## 6.2.3   Morphology

Flax has poorly developed spindle-shaped roots, which reach a depth of 15 to 30 cm. There is a system of lateral roots on the taproot.

Flax stem is straight at the first growth stage and bended with winter plants. The height of the stem from cotyledon to the branching point is 0.5 to 0.8 m. Stems are round, 1 to 2 mm thick, with a waxy surface (Figure 6.2). Its special green color has a cabbage gloss of dark green to light yellow; lighter shades are desired.

The inflorescence of flax is the top part of the stem where side branches appear; this part ranges from 5 to 15 cm in length. Fiber flax is less branched than oil flax, which is in correlation with production purposes. There are 50 to 255 fibers in the stem of fiber flax, while. individual flax fibers are represented by phloem also called bast. Between 30 and 50 fibers are glued into sheaves by pectin. Pith in the middle of the stem dries with maturing and leaves a hole in the stem that is smaller toward the top.

Leaves are sitting, free, alternating, and more or less pointed on top with a slightly waxy surface. The leaf blade is narrow and elongated, with three parallel veins that distinguish it from other dicots.

Flowers are placed on top of stems and side branches. They consist of five petals, five sepals, five stamina, and a pistil. Flax is a self-pollinated species, and the self-pollination ratio is 95 to 98%. Petals are colored and can be white, pink, or blue-violet.

*Figure 6.2* Flax.

The fruit of the flax is a capsule with five segments, each containing two seeds. The flax seed is smooth, chocolate-brown to bright yellow in color, with significant shine. Seeds of some cultivars are yellow or olive green; they are flat, oval; and pointed at one end. The seed is relatively large: 1000-seed weight of fiber flax is 3.4 to 5.4 g, oil (seed) flax is 5.4 to 14 g, and combined flax is 6 to 8 g. The weight of a hectoliter is 65 to 75 kg hl$^{-1}$.

## 6.2.4   Growth and development stages

Flax development stages are as follows: germination, emergence (cotyledon), first pair of true leaves unfolded, third pair of true leaves unfolded (start of leaf spiral), intensive growth (stem extension), bud formation, flowering, and maturing. The intensive growth stage takes between 10 and 16 days. At this stage, daily growth reaches 3.8 cm. Flowering begins 41 to 56 days after emergence and lasts about 1 month, the duration being longest in the earliest emerging crop. The final harvest is carried out once the plants lose all green color. The growth period of spring-sown hemp is 75 (90) to 110 (120) days.

Growing flax for seed demands 100 to 300 m of space isolation, or the plant will be unable to crossbreed.

The stages of flax ripening are as follows: green maturity, early yellow maturity, yellow maturity, and full maturity. During the green maturity stage, the crop seems green although the bottom third of the stem is dry. The grain is still in milky condition.

Early yellow maturity is known for waxy maturity. It bends under the pressure of fingers and gives the impression of wax. Most of the plant is yellow green at this point. Yellow maturity takes place 5 to 7 days after early yellow maturity. Leaves on the bottom half of the stem begin to fall off, and the upper leaves fade. Flax capsules are yellow and partly dark; the grain is hard with a normal light-reddish color.

In full maturity, the capsules and stems darken. Most of the leaves fall off, and the. stems lose their elasticity and become lignified. At this stage, the flax is over-mature.

### 6.2.4.1   Growth conditions

Fiber flax is usually produced as an early variety and oil (seed) flax as a winter variety. Fiber flax flourishes in moderately warm climates where emergence temperatures are optimally between 2 and 5°C) and growth temperatures do not exceed 16 to 17°C. It grows best in moderate to cool conditions, particularly during seed filling, and is thus essentially a crop for temperate regions, because high temperatures and temperature fluctuation at the capsule and flowering stages reduce the number of seeds per capsule, seed weight, oil yield, and quality.

The oil (seed) flax requires higher temperatures than fiber flax and does not tolerate extreme changes in temperature. Young plants of early flax can survive up to –4°C and of winter flax up to –12°C.

Fiber flax belongs to hygrophytic plants with high transpiration coefficients between 400 and 780, or even as high as 1000. Negative drought influence on plants is the strongest from the emergence to the flowering stage. Rainfall or irrigation late in the season can result in a flush of new tillers and leaves, causing uneven ripening [41]. Suitable soil moisture is 70% to feul field soil moisture content. Optimal relative air moisture is between 60 and 70%. Oil (seed) flax also has lower water demands.

The environmental factors most likely to be responsible for yield reductions are high temperature, which causes significant yield losses due to its effect on hastening development rate, and the consequent shortening of the growing cycle and water deficits. The possibility of extending the length of the growing period through earlier sowings seems, however, limited by the difficulties caused by low temperatures in the emergence phase [42].

An opinion on modest light was not confirmed in research because a lack of light reduces photosynthesis and causes lodging. Short intensive growth from the bud formation stage to the flowering stage (50% of nutrient demands in this period) and weak pumping power of flax roots demand favorable nutrient provision. Soil must be airy but, nevertheless, sandy soils and heavy clay are not suitable.

### 6.2.4.2  Cultivation practice

*6.2.4.2.1  Crop rotation.*  Crop rotation is necessary in flax production. It may be sown on the same field only every 5 to 6 years, according to our own experience, though literature on the subject recommends 6 to 11 years between rotations, mostly due to root system diseases. In crop rotation, flax can be sown with all field crops; it is only important that soil be weed free.

Flax growth can be reduced by the allelopathic influence of grasses *Lolium perenne* L. and *Phleum pratense* L, and the consequence of allelopathy is reduced carbohydrate synthesis.

*6.2.4.2.2  Soil.*  Soil for flax production must have a good drainage system: lighter sandy soil to clay soil with a pH between 5 and 7 is suitable. Direct calcification before sowing reduces fiber quality. Very fertile sandy soil is not suitable for flax production even with optimal water supply, because it causes excess stem elongation.

The recommended soil cultivation is a classical one with plowing, harrowing, or presowing machines. Basic presowing soil cultivation for early flax is performed in the usual way for early crops, and for oil flax the same as for winter crops. Due to poor root development, plowing should be performed to full depth. In shallow soil, the sowing layer can be extended to the depth of 20 to 25 cm some years before. Tight soil is loosened. It is important to create exactly the right field surface with presowing cultivation; the sowing layer must have the right structure.

*6.2.4.2.3  Fertilization.*  Due to poor nutrient uptake, nutrients must be available throughout the entire growing period. The highest demand for nitrogen is noted between the  leaf spiral and flower capsule stages; the highest demand for phosphorus between emergence and the leaf spiral stage; and the highest demand for potassium between the first three weeks of growth and the flower bud stage.

In 100 kg of dry matter yield and the corresponding grain amount, between 0 and 1.4 kg N, 0.4 and 0.5 kg P, and 1.1 and 1.6 kg K are found. Between 5000 to 6000 kg stem dry matter and 1000 to 1500 kg seed ha$^{-1}$ can be produced. Combined production reduces seed yield 20 to 30%. In 100 kg of seed, 4 kg N, 1.8 kg P, and 1.15 kg K are found in the soil.

Flax is thus a crop with low nitrogen requirements [43], reducing the risk of nitrate surpluses in groundwater. Fertilization should not be conducted directly with organic fertilizers but only in precrop, in the amount of 20 t of stable manure ha$^{-1}$. The necessary calcification of previous crops should also be conducted.

Special attention is given to nitrogen use. A high content of mineral nitrogen exists in the soil, and abundant fertilization with nitrogen fertilizers can result in lodging, excessive branching, low technological stem yield, and poor yield quality. Winter flax is fertilized in autumn with up to one-third of the joint nitrogen rate, and early flax with one-half of the rate. All phosphorus, potassium, and other nutrients should be added to the soil in presowing.

### 6.2.4.3 Sowing and cultivation practice

Flax grain must have at least 95% germination, a natural shine, and should not be damaged; when rubbed, it should turn fat and purity should be 99%. It should not contain any dodder (*Cuscuta villosa* L.) or ryegrass (*Lolium* spp.) impurities or seeds of cultivars produced for other purposes, such as seeds of oil flax in fiber flax. When oil flax and fiber flax plants cross, the yield quality changes. It cases like this, it is best to buy variety seeds.

Early flax is sown at the same time as early cereals, and winter flax is sown at the end of summer. In Australia, where growth conditions allow autumn sowings, the flax gave higher yields of both stem and seed compared with winter and spring sowings [40].

Flax is sown with the sowing machine for cereals. Fiber flax is sown at the usual interrow spacing of 10 to 12 cm (also 6 cm), oil flax is sown at 20 to 45 cm spacing, and combined cultivars are sown at 15 to 20 cm interrow spacing. The recommended sowing rate of fiber flax is unusually high: 2000 to 3000 seeds $m^{-2}$. Because up to 40% of fiber flax seeds usually fail, the sowing amount of fiber flax is 140 to 150 kg seed $ha^{-1}$. Oil (seed) flax is sown at a lower density; 70 to 80 kg $ha^{-1}$ suffice with the goal of obtaining at least 400, or perhaps between 600 and 700, plants $m^{-2}$. For combined cultivars, 100 to 120 kg $ha^{-1}$ seeds may be needed. In trials, Diepenbrock et al. [41] and Casa et al. [42] found that yield was unaffected by seed rates varying from 200 to 800 seeds $m^2$ in the lowest — as well as the highest — yielding locations. In the other locations, low and high seed rates yielded less than the intermediate rates of 400 and 600 seeds $m^2$ [42]. However, an optimum seeding rate depends on the sowing date, the target production of seed or fiber, the soil quality, and weather conditions, and it involves a compromise between maximizing yield and minimizing potential losses from lodging. In some growing conditions, the crop was able to compensate for reduced stand densities mainly by increasing the number of capsules per plant [44].

The sowing depth should be between 1.5 and 2.0 cm; only occasionally should sowing take place up to 3.0 cm deep in light and dry soil. Crop provision begins by rolling after sowing. In case of crust formation, harrow should be performed, but not after emergence. Flax can be hoed after emergence, at 14-day intervals, when sowing in wide rows.

Organic nitrogen fertilizers are used for additional fertilization of early cultivars one time, three weeks after emergence; this occurs twice for winter cultivars: at the height of 10 cm and at the flowering stage.

In dry conditions, the crop is irrigated. One measure of water should not exceed 30 mm.

## 6.2.5   Harvesting

Fiber flax is harvested at the stage of early yellow maturity, and oil flax at the stage of yellow maturity, the same as combined flax. Seed flax does not require any specialized farm machinery, as sowing and combine harvesting can be carried out with the same equipment as that used for winter cereals. Harvesting of fiber flax can reduce yield due to stubble remains; fiber flax is ready for harvesting when two-thirds of stems are yellow and leaves fall off, which happens a month after the opening of the first flowers. If plants are harvested manually, they can be grabbed without weeds, pulled out, and shaken. They are sorted according to length into three groups: longer than 67.5 cm, 60 to 67.5 cm, and shorter than  60 cm.

Dried plants are put in stacks. Stacks are made of tied sheaves with 15 cm diameter or of untied plants. Heaps are dried for 10 days or until they are completely dry. Dried flax is tied into bigger sheaves, 30 to 40 cm in diameter Special combines are used for pulling out flax. Stems are processed as soon as possible or are stored in extremely dry places.

Oil flax can be harvested by cereal combine, though the plants must be completely dry. Where production allows the use of chemical substances, desiccation is performed when 5% of capsules are still green, and combining is done after a week. Combining speed should not exceed 2 km $h^{-1}$. Ware seed is dried at a temperature of 80°C, and moisture in storage facilities should not exceed 10%; the recommended moisture is between 8 and 10%.

Fiber flax yield can reach between 2500 and 6000 kg $ha^{-1}$. Grain yield usually varies between 220 and 2820 kg $ha^{-1}$; Lisson and Mendham [40] reported that the flax grain yields in trials were between 1560 and 2180 kg $ha^{-1}$. The highest yield of winter crop recorded in the U.S. was 4390 kg grain $ha^{-1}$. Expected yields usually fall between 1000 and 1500 kg grain $ha^{-1}$; combined fiber-grain production achieves yields somewhere in-between.

## 6.2.6   Utilization

Flax fiber represents raw material for textiles (linen and blends with cotton and other fibers) [45]. Flax fiber is suitable for the most beautiful canvasses of the highest quality, such as damasts. Flax fibers are also used for making clothes, underwear, and laces, and as composites, paper [45], and for insulation materials.

Natural flax fiber is a good replacement for glass fiber in automotive parts [46]. The advantages of natural fibers include their lower density, sound absorbance, lower shatter properties than glass, and lower energy costs for producing composites.

The revival of traditional flax canvas production for tourists is another interesting development, and the production of oil flax seed has its own marketing advantages as a delicious supplement for bread and cakes.

All bast fiber plants, including flax, must undergo the process of retting to separate the fiber from the woody cells, which are called "shives" and constitute the major trash component of flax fibers. The retting starts with the dew of dry steams, where decomposition is affected by usual microorganisms or can be focused on enzyme-retting [47, 48] or frost-retting fiber straws in damp air in Finland [49], because the retting process and drying, decomposed nonfiber components must be integrated with the subsequent mechanical processing steps to clean the fiber for specific applications. However, good quality and long fibers could provide a low-cost source of material for diverse applications if harvested and processed without specialized equipment [50, 51]. Short bast fibers are used as insulation and in packaging materials [52]. Shive has been used as a component of packaging materials, but its main use is still as fuel. The properties of raw materials should be studied to enable selection of the right material for appropriate technical applications.

Output efficiency is very important in flax fibers. Stephens [53] reported total fiber yields of straw ranging from 20 to 30% in a series of flax varieties, where the majority of the nonfiber components are shives.

Oils dry very fast and are consequently used for the production of paints, lacquer, linoleum, varnish, ink for printing, herbicides, and many other pharmaceutical or cosmetical products. Flax oil in combination with fiber is a very reliable plumbing seal. It is also used as an additive in baking. Ground seed and oil seed cakes are an important source of nutrition for animals. Flax seed is also gaining its popularity as an additive to breads and cereals. Approximately 100 g of seed contains 6.3 to 6.6% moisture, 18.0 to 20.3% protein, 34.0 to 37.1% fat, 33.6 to 37.2% carbohydrates, 4.8 to 8.8% fiber, 2.4 to 4.5% ash, 170 to 271 mg Ca, 2.7 to 43.8 mg Fe, 0 to 30 μg β-carotene, 0.17 mg thiamine, 0.16 g riboflavin, and 1.4 mg niacin. Raw flax oil contains 0.25% phosphatides, and the content of fatty acids is as follows: 11% palmitic, 11% stearic, 4% hexadecanoic, 34% oleic, 20% linoleic, 17% linolenic, and 3% unsaturated $C_{20-22}$. Amino acids in the seeds are found in the following quantities (g $16g^{-1}$ N): 8.4–10.3 g arginine, 1.5–2.5 g histidine, 2.5–4.55 g lysine, 0.7–1.5 g tryptophan, 5.6 g phenylalanine, 7.50 phenylalanine + tyrosine, 2.3 g methionine, 6.09 g methionine + cystine, 4.37–5.1 g threonine, 6.54–7.0 leucine, 4.0–4.54 isoleucine, and 5.46–7.0 g valine [54, 55]. Ash contains 30.6% $K_2O$, 2.1% $Na_2O$, 8.1% CaO, 14.3% MgO, 1.1% $Fe_2O_3$, 41.5% $P_2O_5$, 2.3% $SO_3$, 0.2% Cl, and 1.2% Si [54].

Flax is a crop that most strongly (compared with cotton and hemp) absorbs and accumulates heavy metals from the soil, suitable for growing in industrially polluted regions. It removes considerable quantities of heavy metals from the soil with its root system and can be used as a potential crop for cleaning the soil from heavy metals [56].

Many positive properties are attributed to flax in folk medicine. It has been used as a laxative and a substance with diuretic, calming, healing, soothing, and anticancer properties. The anticancer substances also contain 3'-dimethylpodophyllotoxin, podophyllotoxin, and β-sitosterol. It is supposed to help with bronchitis, conjunctivitis, and diarrhea. Furthermore, dietary flax seed supplementation could prevent hypercholesterolemia-related heart attacks and strokes [57].

Ground seed is an appreciated compress in folk medicine. It is used also in combination with white mustard and lobelia.

## 6.3   Jerusalem artichoke

### 6.3.1   Introduction and crop description

Jerusalem artichoke (*Helianthus tuberosus* L.; syn.: *H. mollissimus* E. Wats) is a C-3 warm-season crop native to the southern part of North America. It is a member of the family *Asteraceae* and a close-related crop to sunflower (*Helianthus annuus*). In the past, it was used as food for Indians, and in Europe, it was first grown in France. Today it is grown throughout the world, and its utilization has increased slowly over time. The yield is a good nutritional source for diabetics, processing biogas [58], ethanol [59], fructose [59, 60], and substitutes for artificial sweeteners, e.g., inulin [60, 61]. The crop is suitable for organic production, especially due to its well-expressed weed competition and low production inputs.

Jerusalem artichoke forms numerous roots. It is grown mainly for tubers, whose place of formation is underground, where the main root prolongs to the stem. Thickened tubers resemble potato tubers, only their eyes are knotty. Their color depends on variety and includes red, violet, white, ochre, and similar hues.

The aboveground parts of the plants start growing and developing after spring frost, and they join rows in the middle of June. Intensive stem growth takes place between July and the autumn months (depending on the cultivar). The thickening of tubers begins after stems and leaves are fully developed and should continue in winter; therefore, the green parts are not cut until then. The plant can reach a height of 1.5 to 3 m above ground, and stems turn wooden at the flowering stage. The inflorescence diameter is between 5 and 7.5 cm. Late cultivars flower in October and do not produce seeds in continental climate.

The main part of the Jerusalem artichoke used for human consumption is the tuber. It contains about 80% water, and the dry matter (= 100%) contains between 10 and 15% proteins and 75 to 67% extract containing approximately 60% inulin, 1% fat, 4 to 6% fiber, and approximately 5% ash. According to Mullin et al. [62], the Jerusalem artichoke contains 20 to 25% protein on a dry basis and up to 43% dietary fiber. The phosphorus content is around 0.099%, Ca is 0.023%, Fe is 3.4 g 100 g$^{-1}$, and there are some trace elements, such as Al, Cl, I, Mg, K, S, and Zn. The tubers contain a small amount of vitamins B and C, purine-based arginine, histidine, and other compounds like betaine, choline, and hemagglutinins [63].

## 6.3.2   Growth requirements and organic cultivation

Jerusalem artichoke is a particularly interesting and suitable crop for low-requirement environments [64,65]. In spite of some assumptions that the plant is insensitive to dry conditions, its acclimation to water stress strongly reduces growth in the early growth period and aboveground biomass, and marginally modified the fructan's accumulation to the final harvest [66]. It also grows well in humid soil and favors heat as well as colder temperatures. In-soil tubers can resist temperatures as low as –30°C, while aboveground parts cannot survive the frost. The Jerusalem artichoke cannot be grown in very humid soil with poor draining properties. Neutral soil is optimal, but it can also grow in soils where other crops cannot. The variation width for different cultivars is between 4.5 and 8.2 pH.

The Jerusalem artichoke is a prime candidate for organic cultivation. There is no special requirement for crop rotation — the previous crop can include clover, grasses, and clover-grass mixtures. However, grower attention must be focused on soil pests like *Agreotes* sp. and rodent animals, because they can destroy the yield of tubers that will then be unmarketable; the tubers may be completely eaten by an extensive rodent population. The plant can be planted as a fallow crop and grown continuously for 4 to 5 years on the same field. After that, because small tubers remain in the field, weeds can become inconvenient. For this reason, the following crop rotation must include plants for animal feed, which require frequent cutting. It helps to exhaust Jerusalem artichoke, so that the field can be used as a normal arable field. However, the rotation of Jerusalem artichoke like a fallow crop planted before and after grasses, clovers, or its mixtures for 4 to 5 years is suggested.

Most of the culticars, which are extremely varied, originated in Switzerland and France [67]. The cultivars may be separated into two groups: early maturing (such as Grando) and late maturing, such as Kharkov, Miello, Dub, Rico [68], Violet de Rennes, Bianco, and so forth.

Only high yield cultivars are acceptable for wider production with appropriate shape and color. In the U.S., bacteria-resistant cultivars have been created with higher frost tolerance on the green parts, higher pH tolerance, photoperiodic tolerance, and more.

The soil can be previously cultivated, as with potato planting. If necessary, calcification is performed in autumn. At basic plowing, 30 (40) t ha$^{-1}$ of stable manure should be added; it can be replaced by green manure, the plowing of grass-clover mixtures, and so forth. In such cases, green manure or harvest remains can be fertilized with liquid manure. In general, the Jerusalem artichoke does not require a lot of fertilizing; its needs range from 15 to 30 kg ha$^{-1}$ of nitrogen, 40 to 50 kg of phosphorus, and 40 to 60 kg of potassium ha$^{-1}$ per year. After presowing field preparation, Jerusalem artichoke can be planted early in spring or even in autumn.

The suggested growth area is 60 × 40 cm for tuber production and 60 × 30 cm — or even 30 × 30 cm — for green plant mass production. Tubers are

*Figure 6.3* Tubers of Jerusalem artichoke.

planted at a depth of 6 to 10 cm, if seeding depth from the soil top to the tuber is taken into consideration. The quantity of seeding material depends on the tuber size and varies from 1200 to 2000 kg ha$^{-1}$.

During the first year, Jerusalem artichoke is hoed in interrow spacing in mid-spring, when it is shed before covering the rows with aboveground mass. As soon as the soil is covered, weed control is no longer necessary due to intensive overgrowth. In the second and following years, tubers are removed from interrow spacing for easier hoeing the following year. The more extensive approach allows crops to be harrowed only in spring and hoed once more if necessary. Yield is conditioned by soil characteristics and fertilizing. In favorable conditions and appropriate crop provisions, a yield similar to the potato can be achieved; the average yield is between 15 and 30 t of tubers per hectare.

Jerusalem artichoke tubers (Figure 6.3) are not picked for supply but according to the possibility of sale. The best storage place is in soil, except for storage facilities with high humidity (over 80%) and low temperatures between 0 and 5°C.

## 6.3.3   The perspective of the use of Jerusalem artichoke also produced as organic product

Jerusalem artichoke has been an important source of food for the human diet, and it also has medical and industrial applications [69, 70]. It is a recommended food for patients with diabetes, because tubers contain no starch but do contain approximately 15% of inulin. In folk medicine, it was used as a diuretic and aphrodisiac, as well as for stomach problems and other effects. It has been grown to produce fructans that can be used for many purposes, like ethanol production [71, 72] and vinegar.

Jerusalem artichoke is one of the most important candidates for use as a raw material for the industrial production of biological fructose and inulin. Its naturalness makes it more attractive to consumers than synthetic products. The yield of stalks and tubers at flowering time produces a yielding potential of total sugars (fructose + glucose) and inulin from 10.4 to 18.6 and 8.0 to 17.9 t ha$^{-1}$, respectively [73]. Jerusalem artichoke in inulin applications could be used for diagnostic use (inulin with a high degree of polymerization, over 20) and for improving the consistency of cakes and other bakery products (inulin with a smaller degree of polymerization, 6–10) [74]. Fructans can be used in the food industry and in several other industrial and medical applications [66].

Jerusalem artichoke is a welcome feed for organically kept pigs and also a good aboveground mass for the production of snails. It can be grown on an additional open field close to the pigsty, so that the pigs can consume fresh living tubers and young plants. The aboveground biomass is also used as fresh feed or silage (two to three hay harvests per year).

Lignified plants may be used as fibers for natural building materials. When producing for silage, between 30 and 50 t ha$^{-1}$ can be produced, creating a silage mass with a similar starch-protein ratio (1:10) to corn silage. The crop can be cut two to three times in summer and late autumn before frost. When done this way, tuber yield is reduced by half, but higher-quality green fodder is produced.

The plant can also be utilized for biogas production, especially on farms without animals. Anaerobic digestion experiments showed that fresh and ensiled aboveground parts of the plant could produce 480 to 680 liters biogas kg$^{-1}$ organic material [58].

## 6.4   Sweet potato

### 6.4.1   Introduction

Sweet potato, also called batate and ipomea (*Ipomoea batatas* [L.] Lam., Poir., syn.: *Convolvolus tuberosus* Vell., *Batatas edulis* Choisy), originates from Central America; from there, its use spread to other areas and continents. It reached Europe in the same way as beans, potatoes, and corn after Columbus's discovery of America. Sweet potato was carried to Spain for the first time in 1492. The ancestry of the sweet potato is not known due to the inability to gather such information on wild-growing plants. It is a perennial crop but can be produced as annual crop. Although an unknown plant in some temperate climates, it is an important food source in tropical areas. Throughout the world, sweet potato is produced on more than 9 million hectares of land at a total weight of 140 million tons and with an average yield 15,000 kg ha$^{-1}$. The highest amounts are produced in China, the U.S., New Zealand, and Australia; within Europe, it is produced in Italy, Spain, and Portugal across approximately 6000 hectares [75].

## 6.4.2  Botany

Sweet potato belongs to the *Ipomoea* genus and *Convolvulaceae* family. There are many cultivars known in the world, but only in Papua New Guinea do more than 2000 exist (Ivančič, personal communication). The cultivars are produced from vegetative reproduction of selected plants that are a result of natural and artificial crossing. A small number of cultivars are mutated. Sweet potato is a cross-pollinated plant with a strong autoincompatible system, but perfectly autoincompatible genotypes also exist that can be easily self-pollinated.

Seedlings have branching root systems. Adventitious roots appear quickly on the nodes of climbing stems so that the stem is rooted as it grows. Tubers are a result of the secondary thickening of some adventitious roots under the soil surface. When tubers ripen, stems are removed and tubers dug out. Tuber color may be white, yellow, violet, or pink, and they weigh between 0.5 and 2 kg, or even as much as 7 kg with some cultivars (Figure 6.4).

The stem is annual and reaches between 1 and 5 m in length. Stems are partially straight, trailing, or partially trailing. Stem diameter is between 3 and 10 mm, and the internodes are 2 to 10 cm long. Th surface can be smooth or hairy, and the color of the stem is green to purple with occasional violet spots on ground internodes.

Leaves vary according to genotype and age; they are simple and alternating. The first true leaf is usually small, smooth or hairy, cordate, with a pointed or obtuse top. Other leaves are usually larger, measuring 5 to 15 × 5 to 15 cm. Petioles are from 4 to 30 cm long. The leaf surface can be wavy

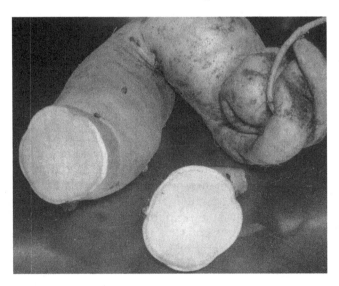

*Figure 6.4* Tuber of sweet potato.

or straight, with two nectar glands at the bottom, and its shape can be whole or palmately lobed. The color is green or violet with occasional violet spots at the bottom. Leaf veins are palmate and possess the same color scheme,

Flowers grow individually from nodes or as inflorescence. A flower consists of five sepals (between 1.0 to 1.5 cm long), five grown petals (funnel-shaped, 2.5 to 5 cm long and 2.5 to 4.5 cm wide), five anthers (grown at the bottom of petals with uneven lengths; filaments are white with gland hair), and a pistil (joined with ovary, with an orange nectar gland). The flower contains two bracts, and petals are white, pink, or violet pink. They open between 9:00 a.m. and 11:00 a.m. in the morning and remain opened in cold or cloudy weather for an even longer time. The fruit is a dark and round pod, usually containing one to two seeds. The seeds are dark and irregularly shape, 3.0 to 4.5 mm long with hard teguments.

### 6.4.2.1  Climatic characteristics, growth, and development

Favorable growing conditions for the sweet potato exist between the geographical coordinates 40°N to 32°S. Successful growth and development are conditioned by 4- to 6-month periods without frost.

The growing stage of the sweet potato lasts between 3 and 12 months, depending on cultivar and climate. Cultivars with longer growing stages than corn (150 to 160 days) should not be chosen in certain areas; in central Slovenia, for example, the growing stage of early cultivars is 4 months. With sweet potatoes, talk is about cuttings and not about seedlings. The main problem with seedlings is their initial slow growth. Emerged plants are ruined by temperature from 0 to 2°C; even developed leaves are not resistant to this temperature but developed stems can survive –2 to –3°C, depending on the development stage, variety characteristics, and soil. Tubers freeze at –2 to –3°C, and soil moisture is important. Tubers most often freeze in sandy soil at lower temperatures than in heavy soil.

In tropical climate, the sweet potato is a perennial plant; in moderate climates, it can be grown as an annual plant. This is the result of temperature demands, because plants stop assimilating at temperatures lower than 10°C. For successful growth and development, temperatures higher than 20°C are required — the optimal temperature is between 30 and 35°C.

The sweet potato plant is cross-pollinated, i.e., insect-pollinated. It rarely flowers in moderate climates, due to longer day lengths; rather, the sweet potato is a short-day plant. Most cultivars will flower in 10- to 11-hour day lengths or 12- to 14-hour night lengths. The plant forms partial flowers at 12-hour day lengths, and flowering stops at 13.5-hour days.

## 6.4.3  Cultivation practice

Sweet potato is less demanding than the common potato (*Solanum tuberosum*). Organically produced sweet potato can be included in all crop rotations but stable manure with precrop is recommended. Stable manure or compost are useful, or even obligatory, in cases of low nutritional requirements and badly structured soils.

It can be produced in heavy clay soil but also grows in sandy soil. In heavy soil, tubers are elongated with crust and less durable; light soil with an appropriate water and air regime is best, and garden or humus soils are not recommended. In such soils, either the primarily vegetative part is developed or thick tubers are not resistant to storage. Basic cultivation is done to the plowing depth. Presowing cultivation and seeding take place when night temperatures go above 10°C. The seeding bed is prepared in the classical way or seeding hills are used. In tropical areas, sweet potato is sown in heaps or high hills because successful growth and development demand airy soil.

The needs for available nutrients before planting sweet potato are as follows: 60 kg N, 90 kg $P_2O_5$, and 90 kg $K_2O$ ha$^{-1}$. The needs for nutrients are higher in the second part of the growing stage. The data regarding nitrogen needs are contradictory, because its surpluses can affect intensive growth of green vegetative mass instead of tubers. Tuber yield goes up with increasing potassium application, but differences exist among varying genotypes [76, 77]. The cultivar maturity group of the sweet potato should also play an important part in nitrogen fertilization recommendations [76, 78]. Hansen [79] and Srikumar and Ockerman [80] found few differences in the chemical composition and "quality" of potatoes subjected to organic or conventional cropping practices. In the field trials, four elements in potato tubers (P, Mg, Na, Mn) and four elements in potato leaves (N, Mg, Fe, B) were influenced by fertilization treatments, while extractable P, Ca, Mg, and Cu were higher in organically fertilized potato plots [81].

Yano and Takaki [82] concluded that the mycorrhizal colonization can influence growth promotion at soil pH 4.2 (twofold increase in whole plant dry weight), but not at pH 5.2. As a result, no significant difference was detected in whole plant dry weight between the mycorrhizal plants at pH 4.2 and nonmycorrhizal plants at pH 5.2. The mycorrhizal plants at pH 4.2 showed reduced toxic symptoms of Mn (brown specks on mature leaves) and Al (poor root growth) compared to nonmycorrhizal ones, but tissue concentrations of P, K, and Ca did not increase in mycorrhizal plants.

The crop is cultivated by hoeing on three to four occasions. Cultivation in the tropics comprises only minimal cultivation right after planting. In West India and the western U.S., occasional cultivation prevents the rooting of trailing stems and consequently the formation of a higher number of small tubers. When soil moisture is below 60% of water capacity, the crop is irrigated.

Sweet potato is harvested when leaves start to turn yellow; this does not apply to continental areas, however, because leaves will remain green until frost due to high moisture and low temperatures. The best technological ripeness indicator is the cut tuber: if it does not turn black when exposed to sun and dries in a half-hour, it is ripe. Unripe tubers secrete sticky, milky white juice. Prior to picking, the aboveground mass may be removed. Mowing is practically impossible, though checking the possibility of aboveground mass silage is advisable. Tubers can be removed with potato harvesting machines or plows, and they should be left to drybefore picking. After harvesting, tubers are storied at 10°C with 75% relative air moisture. Storage taking place 2 to 3 weeks after harvest results in a 10 to 15% loss of tuber weight. Changes in carotene concentration during storage have also been noted [83, 84].

## 6.4.4  Plant reproduction

Reproduction is described in various ways in literature on the subject. According to some sources, plants can be reproduced with seeds or by vegetative means with whole tubers, tuber parts, cuttings, rooted parts of the vine, and tissue cultures. Seed reproduction is used for breeding purposes; for planting at home, small whole or cut tubers and approximately 20-cm-long shoots can be used instead. For wider reproduction, 25- to 40-cm-long cuttings taken from stems, with no roots, may be used.

The plant develops new roots from nodes only when in contact with soil; rooted cuttings can also be grown in protected areas. For 1 hectare, 400 to 500 kg of tubers are needed, from which shoots should be removed and planted for the purpose of developing roots. Plants are ready for planting in 6 weeks, when they will have grown five to six leaves and developed roots. It is simpler to use cuttings taken from stems for this purpose; these are planted in prepared crests in spacings of 70 (100) × 40 (50–100) cm by machine or hand. Sowing material should be sown in soil at two-thirds length. If shoots are used, roots should not dry out. Planting in cloudy weather is most desirable, or the roots should be soaked in a liquid mixture of cattle mud and soil in covered areas.

For simpler production in temperate climates, tubers should be planted in heated, well-protected areas, so that stems reach a length of three m before cutting and planting occurs. Stems are cut into parts with two internodes and planted in crests in light soil. Rooted shoot are rarely used for planting larger areas.

## 6.4.5  Nutritional value and food processing

The quality of sweet tubers varies among cultivars (Figure 6.5). Tubers contain 69 (70)% water, 0.75 (0.2)% fat, 1.8 (3–6)% protein, 26.1% starch and sugar, 1.3% cellulose, and 1.1% ash. The caloric value of 1 kg of tubers is 5154 J. Starch value varies from 10 to 32%, while sugar content reaches 6%. Sweet potatoes grown in the tropics contain more sugar.

In some places, sweet potato is a basic food source, such as bread or potatoes; there are many possibilities for its use. As a vegetable it can be cooked, baked, or roasted. It is also used in sugar, alcohol, and beer production. Sweet potato may be canned and used as food. It is a well-known ingredient in the confectionary industry, and its flour is added when making bread [75]. Sweet potato flour is mainly prepared by drying the peeled slices

*Figure 6.5* Tasting of different sweet potato cultivars.

in a hot air drier, or by drum drying cooked sweet potato mash into flakes, followed by milling and sieving [85]. In addition to the possibility of utilizing sweet potato in wheat-based baked foods, dried and ground sweet potato flour has been investigated as a potential supplement to noodles, puddings, gruel, and so forth [86–89]. Research into the pasting behavior of sweet potato flour obtained by different drying techniques, and the structural properties of its starch, showed that the clustering of starch granules and reduction in their crystallinity as a result of processing decided the properties of sweet potato flour and its suitability to food product development. Sweet potato flour with a low viscosity profile produced by hot-air drying processes is useful in the development of calorie-rich specialty foods and food formulations for children in which a higher solid content per unit volume is required [90]. However, knowledge about changes in carbohydrates during cooking, baking, and the heat moisture treatment of sweet potato [91–94] is essential to determining suitability of sweet potato flour for new requirements, as in snacks, soups, sauces, and more.

## 6.4.6   Health value

The intake of taro, sweet potato, and potato was associated with a decreased risk of kidney cancer death [95]. Anthocyanins from purple sweet potato have antioxidative activity. Results from studies suggest that the antioxidant activity of sweet potato differs depending on plant part and cultivar [96]. However, anthocyanins from purple sweet potato showed stronger 1,1-diphenyl-2-picrylhydrazyl (DPPH) radical-scavenging activity than anthocyanins from red cabbage, grape skin, elderberry, or purple corn, and eight major components of the anthocyanins from PSP showed higher levels of activity than ascorbic acid [97].

# References

1. Jevtić, S., Hemp — *Cannabis sativa* L, in *Posebno Ratarstvo 2*, Jevtić, S., Ed., IRO "Naučna knjiga," Beograd, 1986, 200.
2. Bavec, F., Industrial hemp, in *Nekatere Zapostavljene in/ali Nove Poljščine (Some of Disregarded and/or New Field Crops)*, Univerza v Mariboru, Fakulteta za kmetijstvo, Maribor, 2000, 105.
3. Robinson, R., *The Great Book of Hemp*, Park Street Press, Rochester, VT, 1996.
4. Starčevič, L., Tehnologija gajenja konoplje za vlakno, in *Zbornik Radova "NauNog Skupa Renesansa Konoplje,"* Institut za Ratarstvo i Povrtarstvo, Novi Sad, 1996, 39.
5. Starčević, L. *Hemp (Cannabis sativa L.).*, IFA World Fertilizer Use Manual, Paris, 1992, 477.
6. Nebel, K.M., New processing strategies for hemp, *J. Int. Hemp Assoc.*, 2 (1), 1, 1995.
7. BMLF — Bundesministerium für Land und Forstwirtschaft, *Forschungsbericht 1996*, Vol 23, Abteilung II A1, Ed., Wien, 1997.
8. Lisson, S.N. and Mendham, N.J., Cultivar, sowing date and plant density studies of fibre hemp (*Cannabis sativa* L.) in Tasmania, *Aust. J. Exp. Agric.*, 40 (7), 975, 2000.
9. Callaway, J.C, Hempseed as a nutritional resource: An overview, *Euphytica*, 140 (1–2), 65, 2004.
10. Koivula, M. et al., Emissions from thermal insulations — Part 2: Evaluation of emissions from organic and inorganic insulations, *Build. Environ.*, 40 (6), 803, 2005.
11. Schäfer, T. and Honermeier, B., Effect of sowing date and plant density on the cell morphology of hemp (*Cannabis sativa* L.), *Ind. Crops Prod.*, in press.
12. Amaducci, S., Errani, M., and Venturi, G., Plant population effects on fibre hemp morphology and production, *J. Ind. Hemp.*, 7 (2), 30, 2002.
13. Zitscher, F., *Anwendungen von Geotextilien im Wasserbau (Usage of Geotextiles in Water Building)*, Merkblatt 221, Deutscher Verband Wasserwirtschaft und Kulturbau e.V. (Hrsg.), DVWK-Merkblätter zur Wasserwirtschaft, Verlag Paul Parey, Hamburg, Berlin, 1994.
14. Vogl, C.R. and Heß, J., *Die praktische Hanf Fibel — Informationsbroschüre für den Anbau von Hanf (Cannabis sativa L.) im Biologischen Landbau (The Useful Hemp Book — Information Material for Cultivation of Hemp (Cannabis sativa L.) in Ecologic Agriculture)*, 3rd ed., Druckerei Spörk GmbH, Altenmarkt, Austria, 1997.
15. Glawe, A., in *Funktionelle Agrotextilie für den Schutz der Umwelt (Functional Agrotextile for the Protection of the Environment)*, presented at 3rd International Symposium, Bioresource Hemp and Other Fibre Crops, Wolfsburg, Sept. 13–16, 2000, Nova-Institut, Hrth, 2000.
16. De Groot, B., Hemp pulp and paper production: Paper from hemp woody core, *J. Int. Hemp Assoc.*, 2 (1), 31, 1995.
17. Brunet, J.P. and Lalanne, O., Qualität als Bezahlungskriterium für den Anbau von Faserhanf — Qualitätskontrolle am Beispiel der Aktivitäten der Firma LCDA (Quality as paying criteria for cultivation of fibre hemp — quality control as example the activities of the company LCDA), in *Vortrag zum Projekt Marktinnovation Hanf*, Faserinstitut Bremen, Bremen, 2000.

18. Schmitz, G. und Dämmen, D., Pro Wärme kontra Kälte (Insulation — pro heat contra cold), Öko-Test Sonderheft Energie, 2000, 30.
19. Volmer, M., Mit Hanfdämmstoff in die Baumarktregale (With hemp insulation material in the do-it-yourself store), VDI Nachrichten (News), 11.05.2001.
20. Karus, M. and Kaup, M., Natural fibres in the European automotive industry, *J. Ind. Hemp.*, 7 (1), 119, 2002.
21. Schäfer, D., Einsatz und Potential naturfaserverstärkter Kunststoffe in der Automobilindustrie (Use and potential of natural fibre based plastics in the automobile industry, in *Gülzower Fachgespräche, Nachwachsende Rohstoffe — Von der Forschung zum Markt*, Fachagentur Nachwachsende Rohstoffe, 1998.
22. Schäfer, T. and Honermeier, B., Untersuchungen zum Einfluss des Erntetermins auf den Biomasse- und Faserertrag sowie die Zellstruktur der Sprossachsen von Faserhanf (*Cannabis sativa* L.) (Investigations on the influence of harvest time on biomass and fibre yield as well as cell structure of fibre hemp stems [*Cannabis sativa* L.]), Pflanzenbauwissenschaften, *German J. Agron.*, 2 (7), 92, 2003.
23. Mediavilla, V., Leupin, M., and Keller, A., Influence of the growth stage of industrial hemp on the yield formation in relation to certain fibre quality traits, *Ind. Crops Prod.*, 13 (1), 49, 2001.
24. Keller, A. et al., Influence of the growth stage of industrial hemp on chemical and physical properties of the fibres, *Ind. Crops Prod.*, 13 (1), 35, 2001.
25. Leupin, M., *Enzymatic Degumming Trough Alkalophilic Microorganisms — A New Approach for Bast Fibre Processing. Natural Fibres, Hemp, Flax and other Bast Fibrous*, Plant Production, Technology and Ecology, Institute of Natural Fibres, Poznan, Poland, 1998, 119.
26. Mediavilla, V. et al., Decimal code for growth stages of hemp (*Cannabis sativa* L.), *J. Int. Hemp Assoc.*, 5 (2), 67, 1998.
27. Struik, P.C. et al., Agronomy of fibre hemp (*Cannabis sativa* L.) in Europe, *Ind. Crops Prod.*, 11 (2–3), 107, 2000.
28. Van der Werf, H.M.G. et al., Nitrogen fertilization and row width affect self-thinning and productivity of fibre hemp (*Cannabis sativa* L.), *Field Crops Res.*, 42 (1), 27, 1995.
29. Van der Werf, H.M.G., Wijlhuizen, M., and De Schutter, J.A.A., Plant density and self-thinning affect yield and quality of fibre hemp (*Cannabis sativa* L.), *Field Crops Res.*, 40 (3), 153, 1995.
30. Oomah, B.D. et al., Characteristics of hemp (*Cannabis sativa* L.) seed oil, *Food Chem.*, 76 (1), 33, 2002.
31. Lane, R.H., Qureshi, A.A., and Salser, W.A., Tocotrienols and Tocotrienol-like Compounds and Methods for their Use, U.S. Patent, 6204290, 2001.
32. Jones, K., *Nutritional and Medicinal Guide to Hemp Seed*, Rainforest Botanical Laboratory, Gibsons, BC, Canada, 1995.
33. Pate, D.W., Hemp seed: A valuable food source, in *Advances in Hemp Research*, Ranali, P., Ed., The Haworth Press, Binghamton, NY, 1999, 243.
34. Deferne, J.L. and Pate, D.W., Hemp seed oil: A source of valuable essential fatty acids, *J. Intern. Hemp Assoc.*, 3, 4, 1996.
35. Rausch, P., Verwendung von hanfsamenöl in der kosmetik, in *Bioresource Hemp*, 2nd Ed., Nova-Institute, Cologne, Germany, 1995, 556.
36. McLaughlin, P.J. and Weihrauch, J.L., Vitamin E content of foods, *J. Am. Diet. Assoc.*, 75, 647, 1979.

37. Buchgraber et al., *Produktions — Nischen im Pflanzenbau*, Leopold Stocker Verlag, Graz, 1997, 13.
38. Brown, D., Cannabis: The genus *Cannabis*, in *Medicinal and Aromatic Plants: Industrial Profiles*, Vol. 4, Hardman, R., Ed., Harwood Academic Publishers, OPA, Overseas Publishers Association, 1998.
39. Bavec, F., 2000, Flax, in *Nekatere Zapostavljene in/ali Nove Poljščine (Some of Disregarded and/or New Field Crops)*, Univerza v Mariboru, Fakulteta za kmetijstvo, Maribor, 2000, 55.
40. Lisson, S.N. and Mendham, N.J., Agronomic studies of flax (*Linum usitatissimum* L.) in south-eastern Australia, *Aust. J. Exp. Agric.*, 40 (8), 1101, 2000.
41. Diepenbrock, W. and Iwersen, D., Yield development in linseed (*Linum usitatissimum* L.), *Plant Res. Dev.*, 30, 104, 1989.
42. Casa, R. et al., Environmental effects on linseed (*Linum usitatissimum* L.) yield and growth of flax at different stand densities, *Eur. J. Agron.*, 11 (3–4), 267, 1999.
43. Hocking, P.J., Randall, P.J. and Pinkerton, A., Mineral nutrition of linseed and fiber flax, *Adv. Agron.*, 41, 221, 1987.
44. Diepenbrock, W.A., Léon, J., and Clasen, K., Yielding ability and yield stability of linseed in Central Europe, *Agron. J.*, 87, 84, 1995.
45. Van Dam, J.E.G. et al., *Industrial Fibre Crops: Increased Application of Domestically Produced Plant Fibres in Textiles, Pulp and Paper Production, and Composite Materials*, ATO-DLO, Wageningen, ND, 1994.
46. Lepsch, D. and Horal, J.W., Development of an integrated modular plastic electrical carrier and flax/polypropylene shelf panel for a vehicle rear shelf system, in *Proceedings of the Society for Automotive Engineering International Congress and Exposition*, Society for Automotive Engineering, Warrendale, PA, 1998, Paper #980727, 87.
47. Sharma, H.S.S. and Van Sumere, C.F., *The Biology and Processing of Flax*, Sharma, H.S.S. and Van Sumere, C.F., Eds., M Publications, Belfast, Northern Ireland, 1992.
48. Akin, D.E. et al., Spray enzymatic retting: A new method for processing flax fibers, *Textile Res. J.*, 70, 486, 2000.
49. Kymäläinen, H.R., Technologically indicative properties of straw fractions of flax, linseed (*Linum usitatissimum* L.) and fibre hemp (*Cannabis sativa* L.), *Bioresour. Technol.*, 94 (1), 57, 2004.
50. Foulk, J.A. et al., Flax fiber: Potential for a new crop in the southeast, in *Trends in New Crops and New Uses*, Janick, J. and Whipkey, A., Eds., ASHS Press, Alexandria, VA, 2002, 361.
51. Akin, D.E., Dodd, R.B., and Foulk, J.A., Pilot plant for processing flax fiber, *Ind. Crops Prod.*, 21 (3), 369, 2005.
52. Hautala, M., Pirilä, J., and Pasila, A., *Agro fibre research and industrial development in Finland: Producing strong composites from high quality fibres*, Estonian Agricultural University, Tartu, 2002, Transactions No. 215, 62.
53. Stephens, G.R., Connecticut fiber flax trials 1994–1995, in *The Connecticut Agricultural Experiment Station Bull. 946*, The Connecticut Agricultural Experiment Station, New Haven, CT, 1997.
54. Duke, J.A., *Handbook of energy crops*; Unpublished, Purdue University, New CROP Homepage, 1983, 17 pp.

55. Wanasundara, P.K.J.P.D., Shahidi, F., and Brosnan, M.E., Changes in flax (*Linum usitatissmum*) seed nitrogenous compounds during germination, *Food Chem.*, 65 (3), 289, 1999.

56. Angelova, V. et al., Bio-accumulation and distribution of heavy metals in fibre crops (flax, cotton and hemp), *Ind. Crops Prod.*, 19 (3), 197, 2004.

57. Prasad, K., Dietary flax seed in prevention of hypercholesterolemic atherosclerosis, *Atherosclerosis*, 132 (1), 69, 1997.

58. Gunnarson, S. et al., Jerusalem artichoke (*Helianthus tuberosus* L.) for biogas production, *Biomass*, 7 (2), 85, 1985.

59. Caserta, G. and Cervigni, T., The use of Jerusalem artichoke stalks for the production of fructose or ethanol, *Bioresour. Technol.*, 35 (3), 247, 1991.

60. Stolzenburg, K., Jerusalem artichokes — raw material for inulin and fructose production, *Zuckerindustrie*, 130 (3), 193, 2005.

61. Bemiller, J.N., *Inulin and inulin containing crops, studies in Plant Science*, Vol. 2, Fuchs, A., Ed., Elsevier, Amsterdam, 1994.

62. Mullin, W.J. et al., The macronutrient content of fractions from Jerusalem artichoke tubers (*Helianthus tuberosus*), *Food Chem.*, 51 (3), 263, 1994.

63. Bavec, F., Jerusalem artichoke, in *Nekatere zapostavljene in/ali nove poljščine (Some of Disregarded and/or New Field Crops*), Univerza v Mariboru, Fakulteta za kmetijstvo, Maribor, 2000, 157.

64. Paolini, R. et al., Produttività del topinambur (*Helianthus tuberosus* L.) in relazione a fattori agronomici diversi, *Agric. Ricerca*, 18 (163), 126, 1996.

65. D'egidio, M.G. et al., Production of fructose from cereal stems and polyannual cultures of Jerusalem artichoke, *Ind. Crops Prod.*, 7, 113, 1998.

66. Monti, A., Amaducci, M.T., and Venturi, G., Growth response, leaf gas exchange and fructans accumulation of Jerusalem artichoke (*Helianthus tuberosus* L.) as affected by different water regimes, *Eur. J. Agron.*, 23 (2), 136, 2005.

67. Maillard, A., Techniques culturales et productivite de l'epeautre en Suisse romande, *Revue-Suisse-d'Agriculture*, 26 (2), 77, 1994.

68. enChekroun, M. et al., Comparison of fructose production by 37 cultivars of Jerusalem artichoke (*Helianthus tuberosus* L), *New Zeal. J. Crop Hortic. Sci.*, 24 (1), 115, 1996.

69. Meijer, W.J.M. and Mathijssen, E.W.J.M., Experimental and simulated production of inulin by chicory and Jerusalem artichoke, *Ind. Crop Prod.*, 1, 175, 1993.

70. Harborne, J.B., Inulin and inulin containing crops, in *Studies in Plant Science*, Fuchs, A., Ed., 3rd ed., Elsevier, Amsterdam, 1993.

71. Carrasco, J.E., Experiences on Jerusalem artichoke tuber conversion into ethanol, in *First EEC Workshop on Jerusalem Artichoke*, ECC Report EUR 11855, Luxemburg, 1988, 54.

72. Bajpai, P.K. and Bajpai, P., Cultivation and utilization of Jerusalem artichoke for ethanol, single cell protein, and high-fructose syrup production, *Enzyme and Microb. Technol.*, 13 (4), 359, 1991.

73. Baldini, M. et al., Evaluation of new clones of Jerusalem artichoke (*Helianthus tuberosus* L.) for inulin and sugar yield from stalks and tubers, *Ind. Crops Prod.*, 19 (1), 25, 2004.

74. Vokov, K., Erdelyi, M., and Pichler-Magyar, E., Preparation of pure inulin and various inulin-containing products from Jerusalem artichoke for human consumption and for diagnostic use, in *Inulin and Inulin-Containing Crops*, Fuchs, A., Ed., Elsevier, Amsterdam, 1993, 341.

75. Bavec, F., Ipomea, in *Nekatere zapostavljene in /ali nove poljščine (Some of disregarded and/or new field crops*), Univerza v Mariboru, Fakulteta za kmetijstvo, Maribor, 2000, 109.
76. Warman, P.R. and Havard, K.A., Yield, vitamin and mineral contents of organically and conventionally grown potatoes and sweet corn, *Agric. Ecosyst. Environ.*, 68 (3), 207, 1998.
77. Collins, J.L. and Gurkin, S.U., Effect of storage conditions on quality of sweet potato flour, *Tenn. Farm Home Sci.*, 156, 20, 1990.
78. Emenhiser, C. et al., Packaging preservation of carotene in sweet potato flakes using flexible film and an oxygen absorber, *J. Food Qual.*, 22, 63, 1999.
79. Hansen, H., Comparison of chemical composition and taste of biodynamically and conventionally grown vegetables, *Qual. Plant-Pl. Fds. Hum. Nutr.*, 30, 203, 1981.
80. Srikumar, T.S. and Ockerman, P.A., The effects of fertilization and manuring on the content of some nutrients in potato (var. Provita), *Food Chem.*, 37, 47, 1990.
81. Yadav, R. et al., Changes in characteristics of sweet potato flour prepared by different drying techniques, *Food Sci. Technol.*, 39 (1), 20, 2005.
82. Palomar, L.S. et al., Optimization of a peanut-sweet potato cookie formulation, 27, 314, 1994.
83. Montreka, Y.D. and Adelia, C.B.B., Production and proximate composition of a hydroponic sweet potato flour during extended storage, *J. Food Process. and Preservation*, 27, 153, 2003.
84. Pangloli, P., Collins, J., and Penfield, M.P., Storage conditions affect quality of noodles with added soy flour and sweet potato, *Int. J. of Food Sci. Technol.*, 35, 235, 2000.
85. Chen, Z., Schols, H.A., and Voragen, A.G.J., Starch granule size strongly determines starch noodle processing and noodle quality, *J. Food Sci.*, 68, 1584, 2003.
86. Woolfe, J.A., Post harvest procedures: II. Processing, in *Sweet potato — An Untapped Food Source*, Cambridge University Press, Cambridge, 1992, 292.
87. Damir, A.A., Effect of heat penetration during cooking on some physico-chemical properties and microstructure of sweet potatoes, *Food Chem.*, 34, 41, 1989.
88. Susheelamma, N.S. et al., Studies on sweet potatoes — I, Changes in the carbohydrates during processing, *Staerke*, 45, 163, 1992.
89. Kamolwan, J., Yuthana, P., and Vichai, H., Physicochemical properties of sweet potato flour and starch as affected by blanching and processing, *Staerke*, 55, 258, 2003.
90. Lamberti, M. et al., Starch transformation and structure development in production and reconstitution of potato flakes, *Lebens.-Wiss.+Technol.*, 37, 417, 2004.
91. Ankumah, R.O. et al., The influence of source and timing of nitrogen fertilizers on yield and nitrogen use efficiency of four sweet potato cultivars, *Agric. Ecosyst. Environ.*, 100 (2–3), 201, 2003.
92. Melvin, S.G., Guoquan L., and Weijun Z., Genotypic variation for potassium uptake and utilization efficiency in sweet potato (*Ipomoea batatas* L.), *Field Crops Res.*, 77 (1), 7, 2002.

93. Nin, A. and Gilsanz, J.C., Growth analysis and performance of four sweet potato cultivars under different levels of nitrogen and potassium, *Hort. Sci.*, 33, 443, 1998.

94. Washio, M. et al., Risk factors for kidney cancer in a Japanese population: Findings from the JACC study, *J. Epidemiol.*, 15, 203, 2005.

95. Yano, K. and Takaki, M., Mycorrhizal alleviation of acid soil stress in the sweet potato (*Ipomoea batatas*), *Soil Biol. Biochem.*, 37 (8), 1569, 2005.

96. Kano, N. et al., Antioxidative activity of anthocyanins from purple sweet potato, *Ipomoea batatas* cultivar Ayamurasaki, *Biosci. Biotechnol. Biochem.*, 69 (5), 979, 2005.

97. Boo, H.O. et al., Antioxidant activities of colored sweet potato cultivars by plant parts, *Food Sci. Biotechnol.*, 14 (2), 177, 2005.

# chapter seven

# Legumes

The family *Fabaceae* — commonly known as grain legume species like beans (*Phaseolus* sp.), mung bean, cowpea (*Vigna* sp.), soya bean (*Glycine* sp.), groundnut (*Arachis* sp.), chickpea (*Cicer* sp.), fenugreek (*Trigonella* sp.), and so forth [1] and forage legumes like clovers (*Trifolium* sp.), *Lathyrus* sp., *Vicia* sp., alfalfa (*Medicago* sp.), sarradella (*Ornithopus* sp.), sainfoin (*Onobrychis* sp.), trefoils (*Lotus* sp.), and so forth [2] comprises some 700 genera and 18,000 species [2]. Most of them are adapted to tropical or arid growth conditions [3] and do not include plants of economic interests in temperate and continental regions. However, many species of clovers and other forage legumes — and also some genotypes of grain legumes, like beans, broad beans, peas, lens, soya beans, and chickpeas — are well adapted to temperate and sometimes continental climates.

Legumes are beneficial in crop rotation under organic farming systems. Since the beginning of agriculture, legume crops have been irreplaceable crops for their natural nitrogen fixation abilities. Legumes fix atmospheric nitrogen in a symbiotic association with bacteria of the genus *Rhizobium*, which attaches to the roots and forms nodules. Active nodules can fix up to 600 kg of nitrogen ha$^{-1}$, but a more normal amount is 200 to 250 kg of fixed nitrogen ha$^{-1}$ by forage legumes, and 80 to 150 kg nitrogen of fixed ha$^{-1}$ by grain legumes. For example, in the climbing bean, up to 84% of nitrogen may be derived from fixation [4]. In soils not infected by symbiotic bacteria, the bacteria need to be applied to the seed at sowing. However, absence of bacteria is expected in the low pH soil; in the absence, the host should be planted in crop rotation over many years.

Legumes are frequently used as organic manure, due to their positive effects on soil structure and nutrient availability. Legumes may be also mixed or intercropped by other crops. Important benefits in organic farming of mixed or intercropping cereals with legumes are as follows: efficient competition of cereals with weeds, improved soil structure, reduced loss of plant nutrients, less damage of plants to pathogens and insects [5], and more available nitrogen due to nitrogen fixation. Due to the high content of proteins in the dried grain, grain legumes can be the most important source of proteins for animal feed, because animal proteins are identified as a source of mad cow disease caused by Bovine Spongiform Encephalopathy (BSE). Corn/bean or maize/soybean intercropping may help fix a deficiency of proteins in the silage [6], as well as increased nitrogen digestibility for ruminants compared to corn silage [7].

Legumes play an important role in human nutrition since they are rich sources of protein, calories, certain minerals, and vitamins [8]. According to Iqbal et al. [9], crude protein is provided in the following amounts: lentil 26.1%, green pea 24.9%, chickpea 24.0%, groundnut 26% [10], vigna-cowpea 19–24% [11] to 24.7% [9], and soya bean 24 to 55% [11]. A legume enhances the protein content of cereal-based diets and may improve their nutritional status. Cereal proteins are deficient in certain essential amino acids, particularly lysine [12]. On the other hand, legumes have been reported to contain adequate amounts of lysine, but are deficient in S-containing amino acids (methionine, cystine, and cysteine) [13]. Amjad Iqbal et al. [9] conclude that legumes like chickpeas, lentils, cowpeas, and green peas are rich in lysine, leucine, and arginine and can fulfill the essential amino acid requirement of the human diet, except for S-containing amino acids and tryptophan. In order to compensate for the deficiency of certain essential amino acids in legume protein, they must be supplemented with other vegetables, dairy products, or meat.

The high value of proteins and rich amount of essential amino acids in its seeds rank grain legumes among the top sources of human nutrition throughout the world. Legume proteins can help people avoid a serious nutritional problem, called protein calorie malnutrition, especially among children in developing countries. In this book, however, only grain legumes that are more or less suitable for growth in temperate climates and their modest applications to organic farming practice will be covered.

## 7.1  Chickpea

### 7.1.1  Introduction

The chickpea, also called garbanzo, bengal gram, or egyptian pea (*Cicer arietinum* L., syn.: *C. edessanum* Stapf., *C. grossum* Salisb., *C. sativum* Schkuhr) is the third most widely spread grain legume in the world, after beans and soya beans. The chickpea is produced in the Asian part of Turkey and India; other important producers include France and Spain [14]. The lack of organic chickpea on the world market is evident, and it is mainly certificated by the company Organic India. It is frequently used in vegetarian dishes.

Chickpea seeds need a longer time for cooking than bean and lens. The seeds contain 13 to 32% of proteins, which are similar to animal proteins. The percentage of eight essential amino acids (especially arginine, lysine, threonine, valine, and phenylalanine) is especially important; the seeds contain 1.5 to 5% of cellulose and 2 to 5% of ash. The percentage of oil is relatively high (2.7%) in comparison to other legumes; only the grains of soybeans are higher. The chickpea seed is rich with essential phosphorus compounds, provitamin A, and vitamin $B_1$ [14].

Intensified secretions of the pancreas may be attributed to consumption of the chickpea, and it is also a known diuretic due to the content of asparagic acid in its seeds.

## 7.1.2 Botany

Chickpea is a member of genus *Cicer* L. and family *Fabaceae* (*Leguminosae*). Genus *Cicer* is one of 27 species — 22 perennial and 5 annual [15]; only thechickpea (*Cicer arietinum* L.; arietum-men ram head) is cultivated. There are four subspecies: *Cicer arietinum* L. ssp. *orientale* Pop., *Cicer arietinum* L. ssp. *asiaticum* Pop., *Cicer arietinum* L.ssp. *euroasiaticum* Pop. Ans., and *Cicer arietinum* L. ssp. *mediterraneum* Pop. All are divided into numerous genotypes.

Chickpea is an annual crop, with tuff and roots that extend approximately 1 mm deep. The roots, which form numerous nodules, can be affected by nitrogen-symbiotic bacteria from the genus *Rhizobium*. The plant may be erect, with numerous branches, and the lower part of the stem lignifies at full maturity. The pilose stems grow to lengths of 0.2 to 0.5 m, sometimes as much as 1.0 m. The first two leaves on the stem are scale-like, and further leaves are pinnate with 9 to 17 leaflets and a midrib terminating with a leaflet. The leaflets are about 8 to 18 mm long and 3 to 10 mm wide, and the complete leaves reach about 5 cm in length. The leaflets are elliptical, ovate, or obovate, with serrated margins. The flowers may be white, red-blue, or pink, and quite small (about 1 cm, sometimes to 2.2 cm). Individual flower clusters are auxiliary. The flowers are hermaphroditic, usually solitary, and born on peduncles 2 to 4 cm in length. The number of flowers per plant varies from 40 to 100, depending on climatic conditions, cultivar, and cultivation practice. The fruits are swollen and oblong pods (Figure 7.1). The pods of light colored seeds are straw yellow; the pods of dark colored seeds are ash violet. The pods grow between 1.5 and 3.0 cm long, contain only one or two seeds, and may be sterile at times. Seeds vary in shape from angular to round, and from smooth to very wrinkled; seed color may be white, yellow, orange, red-brown, or black. Comparing 1000-seed weight, the chickpea is separated into three groups: below 200 g, between 200 and 300 g, and above 300 g.

## 7.1.3 Climatic conditions, growth, and development

Chickpea needs a sum of average temperatures between 1800 and 2000°C. This annual crop is partly tolerant to cool temperatures; it can be grown in all regions where the climatic conditions are suitable for vine growth. Compared to other grain legumes, it is the most cool-weather-tolerant crop, except for the pea. The minimum temperature for germination is 2 to 3°C, but the optimum is between 20 to 25°C [15]. Germination under cold conditions

*Figure 7.1* Chickpea plant with fruit.

often results in the decay of germinating seeds. Young plants can tolerate temperatures as low as −16°C after emergence.

Chilling (< 15°C) during the reproductive phase of the chickpea leads to the loss of flowers and pods, infertile pods, smaller seeds, reduced seed yields [16], and negative effects on the functions of reproductive structures [17]. In cold-stressed plants, seed number per 100 pods, seed weight per plant, average seed weight, and average seed size decreased by 35, 43, 41, and 24%, respectively. Seed reserves of starch, protein, and fat decreased by 34, 33, and 43% respectively, while total soluble sugars increased twofold. The accumulation of proteins such as globulins and albumins was inhibited to a greater extent than that of prolamins and glutelins. Most of the amino acids decreased as a result of stress, while some — such as proline and glutamic acid — increased significantly [16].

The young plants are also tolerant to drought; the chickpea transpiration coefficient is lower than most other crops. In the case of high relative humidity, the plants lose flower buds. However, full irrigation in a cool to temperate, sub-humid climate from emergence to physiological maturity always gives the highest seed yield (> 4.7 t/ha), and there is no indication of a critical period of sensitivity to water stress [18].

The bud formation starts in long-day circumstances. The duration of the growth period is between 70 and 100 days, depending on daylight, humidity, and temperature. In the case of high temperatures, the growth period is shorter. The plants grow very slowly for the first 14 days; after that, growth and development speed up. The stems are thin, characteristic of xerophytes plants (small plants with small masses of leaves; the plant is pilose). The flowering stage lasts about 20 days, and the seeds are mature about 40 days afterward. Chickpea flowers are mainly self-pollinated. The pollen and pistil matured 1 to 4 days before the flower is open. Only under extreme drought periods may the 3 to 5% of flower be cross-pollinated. All of the pods mature in each period. The senescence is intensive, causing the leaves to fall away from the plant, but the pods remain strongly attached to the stems. At the stage of over-maturity, the pods fall away, but they do not lose the seeds. The pods breaks after one week, and many grains may be lost during harvesting.

## 7.1.4 Cultivation practice

The chickpea is a very appropriate crop for organic production. It does not need a special previous crop in crop rotation, but there are legumes that should not follow other legumes. The best previous crops are potato and sunflower. Chickpea grown in monoculture results in lower yielding; it is a smart choice for a previous crop when the nitrogen needs of following crops are a bit higher.

Chickpea is the crop of calcareous and marly soils, grows well in sandy soils, and tolerates salt soils better than most other crops. Acid and heavy soils, conversely, are not suitable for chickpea production. Slightly acid soils must be calcificated at the soil tillage in autumn, a year before the chickpea is sown. The most common soil tillage for chickpea is the same as for other spring crops.

Cultivars are known from Hungary (Pax, Kompolti bordo), Greece (Euros, Gravia, and Thiva), Spain (Candil, Castud, and Tizon), and Canada (Oxley), and there are 12 more cultivars from Maroko and 9 from Australia.

With a yield of 2000 kg grain ha$^{-1}$, the uptake of nutrients amounts to 100 kg of nitrogen, 22 kg of phosphorus, 45 kg of potassium, and 33 kg of calcium ha$^{-1}$. Despite the possibility of nitrogen fixation, nitrogen must be available at the young stages. Fertilization with composted stable manure or other organic fertilizers with at least 30 to 40 kg nitrogen ha$^{-1}$ is recommended. Top dressing is also necessary if a lack of available nitrogen is noted in the soil and the plants are without nodules inoculated by *Rhizobium*; 20 nodules per plant are required for symbiotic fixation of nitrogen to be considered effective enough. The phosphorus and potassium fertilizers must be added according to the deficits in the soil by allowed fertilizers. In the case of sandy soils, the fertilization with stable manure from 15 to 30 t ha$^{-1}$ may affect resources of available nutrients, especially available nitrogen from the potential mineralization process.

Inoculation of biofertilizers (vesicular arbuscular mycorrhiza and phosphate-solubilizing bacteria) significantly increased yield attributes of pods/ plants and seeds/plants and yield of chickpea. Among biofertilizers, dual inoculation of vesicular arbuscular mycorrhiza and phosphate-solubilizing bacteria markedly enhanced these yield attributes and yield compared with vesicular arbuscular mycorrhiza or phosphate-solubilizing bacteria alone [19]. Rudresh et al. [20] also report that the effect of a combined inoculation of *Rhizobium*, a phosphate-solubilizing *Bacillus megaterium* sub sp. *phospaticum* strain-PB and a biocontrol fungus *Trichoderma* spp. increased germination, nutrient uptake, plant height, number of branches, nodulation, pea yield, and total biomass of chickpea compared to either individual inoculations or an uninoculated control. Vesicular arbuscular mycorrhizal infectivity was also improved significantly with combined inoculants and farmyard manure compared to sole application of farm yard manure [21].

The deficiency of molybdenum may reduce the percentage of proteins and yielding [22]. According to Nautiyal et al. [23], both deficiency and excess of molybdenum deteriorated the quality of seeds by increasing the content of phenols, cysteine, and albumin and decreasing that of methionine, lysine, legumin, and vicilin protein fractions, apart from reducing the seed weight.

The old seed is not suggested for sowing because the seeds lose germination very fast. The seeds must be healthy, with high 1000-seed weight. If the chickpea is grown for the first time in the field, the seeds need to be inoculated with symbiotic bacteria. Organic seeds can decay in unsuitable climatic conditions; the plants can slowly emerge at 4 to 6°C. Therefore, it is possible to sow on a relative early sowing date, but sufficient soil moisture can increase the seed decay. For orientation, chickpea sowing data can be about 2 weeks before the usual date for sowing corn.

Chickpea can be sown in interrow spacing from 40 to 45 cm or even 50 cm, when interrow hoeing or other accepted weeding for organic farming is used. In case of narrow inter-row spacing of 15 to 20 cm or 30 to 35 cm, mechanized hoeing is not possible and not acceptable for organic farming in large farms. The suggested plant population is 35 to 40 germinated seeds $m^{-2}$ at wide interrow spacing and 50 to 60 viable seeds $m^{-2}$ at narrow interrow spacing; the sowing rate is 70 to 100 kg seeds $ha^{-1}$ for wide interrow spacing and 100 to 150 kg seeds $ha^{-1}$ at narrow interrow spacing where 1000-grain mass is between 200 and 250 g. For higher 1000-grain mass, the seeding rate must also be higher.

Crust soils must be hoed carefully, without damaging the plants. After emergence, the plants need to be hoed twice: first when the plants are 10 cm high, and second at the stage of flower buds. In organic farming, weeding with a harrow comb is suggested, and the best effect can be achieved at midday, when the plant turgor is lower.

Chickpea must be kept weed-free between the five-leaf and full-flowering stages (24–48 days after crop emergence) and from the four-leaf to beginning-of-flowering stages (17–49 days after crop emergence) at the two sites, respectively, in order to prevent > 10% seed yield loss. At both sites, reduction in seed yield, because of the increased weed interference period, is accompanied by simultaneous reduction in plant dry weight, number of branches, pods per plant, and 100-seed weight [24, 25].

In hot climates, soil solarization is a preplanting technique used to control weeds and soil-borne pathogens consisting of mulching the soil surface with polyethylene sheets. On the basis of beneficial effects (improved grain yield, increased nitrogen availability in soil, nitrogen accumulation in plants, and improved plant growth), soil solarization — which avoids site contamination and is suited to organic farming — should be a good opportunity in

Mediterranean areas where the level and stability of grain yields are low and the infestation of parasitic weed *Orobanche crenata* is high [26].

The maturity of the crop is uniform in 3 to 5 days. Because the plants mature very quickly, the rapidly approaching harvest time is more or less a surprise. The pods fall off the plant, especially in rainy periods and at the stage of over-maturity, and the combine must be adapted. By manual harvesting, the plants have to be cut and not pulled, in order to preserve the symbiotic bacteria in the soil. The pods can be shelled like beans.

### 7.1.5 Some remarks on the nutritional value of chickpea and other grain legumes

With 24% crude proteins in the chickpea seeds, the essential amino acid composition of chickpea is as follows: arginine 8.3%, histidine 3.0%, isoleucine 4.8%, leucine 8.7%, lysine 7.2%, methionine 1.1%, phenylalanine 5.5%, threonine 3.1%, trypthophane 3.1%, and valine 4.6% [27]. However, ordinary cooking resulted in the improvement of protein and starch digestibility of the food legumes by 86.0–93.3% and 84.0–90.4%. The reduction in the levels of nutrients is also significant, along with an improvement in protein and starch digestibility, respectively [28], and content of saponins, which was observed after cooking food legumes. Despite the high content of amino acids in the seeds, cooking decreased contents like lysine, histidine, and arginine. In addition, protein solubility and vitamin content decreased with increased cooking time. Chickpea should be cooked in an autoclave at 121°C for no longer than 1 hour to minimize losses in vitamins and amino acids [29].

## 7.2 Groundnut

### 7.2.1 Introduction

Groundnut, also known as peanut or goober (*Arachis hypogaea* L., syn.: *A. africana* Lour., *A. americana* Tenore, *Arachnida hypogaea* Moench) is a subtropical crop, mainly grown from the equator to 40° of geographical longitude. It is cultivated in 107 countries; annual production in 2001 was 35.09 million tonnes from 25.54 million hectares [30]. The groundnut crop is cultivated in 27 Asian countries, accounting for 67% of global production and 58% of global area. The two major producers in Asia are India, with 8.2 million hectares (55.9% of Asia), and China, with 4.6 million hectares (31.6% of Asia). Average productivity in Asia (1.6 t ha$^1$) is higher than the world average (1.37 t ha$^1$). Groundnut is an important worldwide source of food, protein, fat, minerals, and vitamins in the diets of rural populations, especially children. It is the most important major crop in all of India. Of its total production, 12% is for seeds, 8% for edible purposes, 70% for extraction of oils, and 10% for export; hence, it is considered an economically important crop [31]. Within Europe, it is grown in Spain, Italy, and other Mediterranean regions. Groundnut also experienced some efficient growth periods in temperate climates, as in Vojvodina, but its production strongly depends on climatic conditions. Thus, in temperate regions, production may be limited. An average yielding varies but may reach 860 kg ha$^{-1}$ in Asia, 743 kg ha$^{-1}$ in Africa, 1230 kg ha$^{-1}$ in South America, 2200 kg ha$^{-1}$ in Europe, and 2560 kg ha$^{-1}$ in the U.S. [32]. The biology, processing, and utilization of groundnut are extensively described [33] and investigated, but a great lack of research and description of organic farming methods exists for this crop.

## 7.2.2   *Climate, growth, and development*

As a subtropical plant, groundnut needs an average sum of 3000°C for its vegetation period. It emerges at a temperature between (12)14 and 15°C; for flowering, a temperature above 20°C is required. The optimal temperature is a middle temperature, around 23.5°C, and active assimilation takes place within the range of 10.5 to 28.5°C. Plants flower after 30 and sometimes more than 40 days. Leaves are easily destroyed by frost; groundnut grows and develops most successfully at warm temperatures. It demands water but can survive short periods of drought, and late autumn rainfall causes seed emergence in soil. The growing stage takes from 90 to 150 days, and sometimes longer.

## 7.2.3   *Morphology*

The groundnut plant is similar to low bans, though the leaves resemble clover leaves (but evenly divided) (Figure 7.1). Loose roots are placed in spiral rows around lateral roots. Roots are branched and usually have some bacteria nodules. Long-trailing stems growing 40 to 80 cm, with opposite leaves similar to broad bean leaves, sprout from every direction. Leaves are hairy on the bottom and flowers are joined in inflorescence growing from nodes. The calyx is intensely yellow; pistil ovary is placed at the bottom flower part near the node. The ovary continues into a long style, reaching through hypanthium, and ends with stigma. After successful fertilization, an elongated peg carpophore bearing ovules is formed; it bends to the ground and penetrates the soil. At a depth of 6 to 10 cm, it forms a pod containing two or more seeds resembling hazelnuts. Each pod grows to a length of 10 cm. The 1000-seed weight is 200 to 250 g, and the hectoliter weight is 70 to 80 kg hl$^{-1}$ [32; Ivanicič, personal communication].

## 7.2.4   *Genotypes*

The *Arachis* genus consists of 70 species, among which only groundnut (*Arachis hypogaea*) is suitable for production.

   The groundnut (*Arachis hypogaea*) core collection for Asia consists of 504 accessions, 274 of whice belong to subspecies *fastigiata* (var. *fastigiata* and *vulgaris*) and 230 to subspecies *hypogaea* (var. *hypogaea*). The *hypogaea* group takes longer to flower, has more primary branches, longer primary and cotyledonary branches, more nodes on cotyledonary branches, more total pods, mature pods, and pegs per plant, longer and wider pods, and heavier seeds than the *fastigiata* group [34*].

Old cultivars appropriate for production in the Slovenian Primorska region are panski beli, Amarelo, and Valencija. In Hungary, the cultivars Makó (Kecskemet) and Kiszombori (Szeged) have been registered since 1986 and in Greece, the tetraploids Lakonia 4x and Seraiki.

Cultivars of the Virginia type, with alternate branching seeds, have a longer dormancy stage. There is little or no expressed seed dormancy in the subsequent branching of Spanish Valencia.

In the U.S., the number of registered groundnut cultivars has increased over the years (Virginia 81 from 1982; NC 6, NC 7, and NC-VII from Raleigh, North Carolina, 1998). In addition to cultivars from Oklahoma, Florida and Virginia, the number of cultivars from Asia (Icrisat), Indija, and Ghana (Sinkazei) also increased. The register of OECD also includes cultivars from Argentina (Florman and Manfredi) and North America (Selie). The cultivar Dixie Giant in the U.S. is the major germplasm source in all pedigrees of runner market-type groundnuts, while Small White Spanish-1 cultivar exists in 90% or more pedigrees. These two lines contributed nearly 50% of the germplasm of runner cultivars [35].

## 7.2.5 Cultivation practice

Structured, airy, and humus soils are suitable for groundnut. Unstructured soil easy that easily forms crust is unsuitable. It will grow in acid and basic soil but favors pH 6.5. In crop rotation, groundnut is sown after cereals, and It is sufficient for the previous crop to leave behind light soil as long as enough calcium is available. Without enough calcium, plants develop slowly, form poor flowers, and have lower yield that is more affected by disease.

The optimal sowing date is similar to that for corn. The seeds can be sown in heaps with few seeds together or at interrow spacing of 50 to 60 cm or 70 to 80 cm. For sowing, 60 to 90 kg seed ha$^{-1}$ are required to reach 250,000 plants ha$^{-1}$, and the sowing depth should be 5 cm. The groundnut can be also intercropped [36–38]. The maximum recorded monetary advantage was for the groundnut corn intercropping system in the semiarid tropics of India. There was a 20% reduction in nodule mass in intercropped groundnuts in association with corn. Yield advantage in terms of Land Equivalent Ratio (LER) was greatest (1.68) in the groundnut/corn association [39].

Healthy organic fertilizers are recommended in autumn with calculated incorporation of 80 to 100 kg $K_2O$ ha$^{-1}$ and 80-100 kg $P_2O_5$ ha$^{-1}$. Before sowing, nitrogen needs amount to about 30 kg N ha$^{-1}$ if the symbiotic processes will start actively. Rao and Shaktawat [40] reported that application of farm-yard manure 10 tones ha$^{-1}$ and poultry manure 5 tones ha$^{-1}$ under rainfed

* Reprinted from *Field Crops Res.*, 84 (3), Mallikarjuna Swamy, B.P. et al., Phenotypic variation for agronomic characteristics in a groundnut core collection for Asia, 359, Copyright 2005, with permission from Elsevier.

condition increased pod yield of groundnut by a mean of 14.0 and 11.3%, owing to farmyard manure and poultry manure application, respectively, over the control (1620 kg ha$^{-1}$). *Bradyrhizobium* inoculation and mycorrhiza (*Glomus* spp.) statistically increased yield, 1000-seed weight, hydration coefficient, and content of P, K, Na, Cu, and Mg. Use of bacteria and mycorrhiza is cheap and simple in organic farming and it influences development as well as height and quality of yield [41].

Right after emergence, when plants have grown two pairs of leaves, hoeing should be done for the first time; this procedure should be repeated two to three times in 14-day intervals. With first hoeing, the plants are thinned to just two to three plants if sowing was done in heaps. Regular shedding should be performed in order to enable the plant to form a high quantity of pods. Irrigation and top dressing are usually implemented at the beginning of fthe lowering stage and during pod formation. Plant protection is not simple, because the groundnut is threatened by many diseases and pests; however, the crop was not threatened by them in the nontraditional groundnut-growing regions of Vojvodina and Serbia in 1995. The following year, production was threatened by unfavorable climatic conditions.

Groundnut is mature when leaves turn yellow and grains in pods color. At this stage, the plants need to be plowed or pulled. The plants should be left on the ground to dry for a couple of days; later, they will be placed on drying equipment or hayracks. Pods are torn off of dry plants and cleaned in special separation machines. Expected yield in organic farming varies between 1500 and 3500 kg seeds ha$^{-1}$ and can be lower and higher, depending on genotype adaptability and numerous growth factors.

## 7.2.6  Nutritional value and use

Groundnut is first of all an oil crop, because its seeds contain 45 to 50% of quality oil. After pressing, the seed cakes maintain about 45 to 60% of crude proteins; the seeds contain about 26% of crude proteins. Approximately 100 g of shelled crude seeds contain more than 2093 J of energy, 4 to 13 g of water, 21.0 to 36.4 g of protein, 35.8 to 54.2 of fat, 6.0 to 24.9 g of carbohydrates, 1.2 to 4.3 g of fiber, 1.8 to 3.1 g of ash, 49 mg of Ca, 409 mg of P, 3.8 mg of Fe, 15 µg β-carotene equivalent, 0.79 mg of thiamin, 0.14 mg of riboflavin, 15.5 mg of niacin, and 1 mg ascorbic acid [42*].

Raw or roasted seeds sold as "kikiriki" are popular in temperate regions. In Asia and the U.S., the groundnut is used in many ways. It is used for oil, margarine, and peanut butter production; it is also canned, fried, added to cakes and breads, soups, cold cream, pharmaceutical products, ointments, emulsions for insect protection, and fuel in diesel motors.

Groundnut is consumed in boiled or roasted form, and also as a groundnut cake ("kulikuli"). The dry roasted groundnut snack is, at present, the most widely consumed form. It can be consumed alone or combined with

---

* Reprinted from Vasilevski, G., Nout - *Arachis hypogaea* (L.), in *Posebno ratarstvo*, 1. Part, Jevti, S., Ed., 388, Copyright (2006), with permission from Nauna knjiga.

dry roasted corn (popcorn), "gari," coconut, bread, or plantains. The traditional method of preparing dry roasted groundnut first involves sorting out the physically damaged and moldy kernels of raw groundnuts, followed by soaking them in water for about 20 minutes, salting with sodium chloride to taste, and then roasting by stirring the kernels in hot sand placed in an earthenware pot on an open fire. Upon cooling, roasted groundnuts are separated from the sand by a metal sieve [43]. Not much has been left in writing about groundnuts in folk medicine. It was thought to possess aphrodisiac, haemostatic, and protective hemophilic elements, along with other effects.

### 7.2.7 Dangerous aflatoxins

The seeds are susceptible to attack by a number of fungal diseases (the most well known is *Aspergilus flavus*). Closely related secondary fungal metabolites are aflatoxins, which are very powerful hepatocarcinogens, and naturally occurring mixtures of aflatoxins have been classified as the most dangerous human carcinogens [44]. Aflatoxins are a group of highly oxygenated heterocyclic compounds with closely related structure of $B_1$, $B_2$, $G_1$, and $G_2$. Aflatoxin synergizes with other agents, such as hepatitis B, in the causation of liver cancer [45]; it has been linked with the high incidence of liver cancer in Africa [46]. The $LD_{50}$ for aflatoxin-sensitive organisms is 1 mg aflatoxin per $kg^{-1}$ body weight. Carcinogenic effects are present with the use of 15 g aflatoxin $kg^{-1}$ food. The $B_1$ form of aflatoxin shows hematocarcinogenic effects [32]. In Nigeria, aflatoxin $B_1$ was found in 64.2% of samples. Aflatoxins $B_2$, $G_1$, and $G_2$ were detected in 26.4%, 11.3%, and 2.8% of the samples, respectively, in the contaminated samples [43]. Producers and consumers of conventionally and organically grown groundnuts must be aware of these potential hazards.

## 7.3 Soybean: small attention to the important crop

### 7.3.1 Introduction into utilization

Production of soybean, also called soya bean, and soy (*Glycine max* [L.] Merr., syn.: *G. hispida* [Moench] Maxim., *Soja max* Piper) grew in China over 5000 years ago and spread through Asia quite early. Its production began in Europe during the nineteenth century. Soybean is produced in the highest quantities in the U.S., Asia (mostly China), Europe, and Africa. It is one of the most extensively described and researched plants; therefore, we would like to pay special attention to its organic production and importance. The use of organic soybean products is increasing, but more research regarding the organic farming of soy still needs to be conducted.

Soy is one of the most important plants for protein production in the world; its seed contains 24 to 55% protein. It is also an important oilseed, with 17 to 24% oil content. Its fat consists mostly of polyunsaturated fatty acids (5 to 10% linoleic and 43 to 57% linolenic). The carbohydrates content is quite low (2.7 to 12%); carbohydrates are comprised of 5.2% galactose, 5.4% pentose, 3.6% saccharose, 1.3% raffinose, 0.8% invert sugar, 0.4% starch, 2.8% cellulose, and approximately 0.4% dextrins. Ash content is between 3.7 and 5.9% [47]. Soy a semidrying oil with an iodine value between 85 and 130 and is mostly used raw for its protein content. Additionally, soy oils or foods taste like fish after frying due to their thermal treatment. The soybean comprises an important part of animal nutrition. Soy cakes that remain after oil pressing, flour, and groats can be used as high-quality protein feed. Soy as a protein source has been gaining importance in organic farming, organic breeding, and anywhere that animal proteins (the cause of mad cow disease) or genetically manipulated plants have to be replaced with less controversial proteins grown according to organic farming guidelines. Whole plants provide quality feed for silage, most often prepared in combination with corn, resulting in a favorable starch-protein ratio. Whole soy plant contains approximately 15.1% protein, 45.5% nonnitrogen extract, and 11.1% mineral substances.

People use the seeds as a main food or side dish in the human diet, and also in flour, soy milk, cheese, and butter. Soy is an important raw material in food processing (bread, pasta, margarine, grease) and the chemical processing industry (plastic material, glue, insulation material, rubber, soap, oil colors). Besides its protein and oil, the high vitamin content of soy is also important; it contains substantial amounts of the provitamins A, $B_1$, $B_2$, C, D, E, and K. As a legume, it requires between 120 and 300 kg of nitrogen ha$^{-1}$ in soil. The nutritional and biological values of soy used as food have been stressed lately, especially for people who consume little or no animal proteins. The biological protein value in soy seeds is much higher than other proteins in the human and animal diet (Table 7.1); 0.6 to 0.8 g kg$^{-1}$ of soy protein equal 0.4 g of egg protein.

Soy combined with cereals can cover most essential amino acid requirements in a vegetarian diet, satisfying the demands for cystine and methionine, (present in low amounts in soy) and lysine (present in low amounts in cereals). The amount of methionine present in meat cannot be replaced this way, however. Despite the high biological value of soy proteins, their digestibility is worse than other proteins due to the presence of two substances that hinder the digestion of proteins, trypsin inhibitor and hemagglutine. Cooking for long periods of time improves the digestibility of soy proteins but decreases their biological value. Studies have shown that children who consume unbalanced vegetarian diets without the benefit of industrially synthesized additives in soy products do not develop as well as children who consume meat and milk. Soy consumption reduces absorption elements, such as iron and zinc. Some people simply cannot tolerate soy for medical reasons.

Nutritionists and doctors attribute more positive effects to semisaturated soy fats, favorable amino acid rations, and other grain substances. These

*Table 7.1* Amino Acids in Soyabean Comparing Some Important Products [2]

| Amino Acid | Soyabean | Wheat | Beef | Milk | Eggs |
|---|---|---|---|---|---|
| | | Kg/16 kg | Nitrogen | | |
| Arginine | 7.0 | 3.6 | 5.3–5.5 | 3.7 | 5.7 |
| Cysteine | 1.2 | 4.0 | 1.4 | 0.8 | 2.3 |
| Phenylanaline | 5.3 | 4.9 | 3.8–4.5 | 5.0 | 5.1 |
| Histidine | 2.5 | 1.3 | 3.7–3.9 | 3.2 | 2.3 |
| Izoleucine | 6.3 | – | 5.2 | 5.6 | 6.8 |
| Leucine | – | 9.3 | 8.1–8.7 | 9.8 | 9.2 |
| Lizyne | 6.6 | 2.5 | 9.2–9.4 | 7.9 | 7.4 |
| Methionine | 1.2 | 0.7 | 4.1–4.5 | 2.7 | 3.0 |
| Threonine | 4.0 | 2.5 | 4.8 | 5.1 | 4.1 |
| Thrypthopane | 1.1 | 1.2 | – | – | 1.0 |
| Valine | 6.7 | 2.9 | 4.8–5.5 | 6.1 | 6.9 |

*Source*: Reprinted from Pokorn, D., *Forma 7*, Ljubljana, 11. Copyright 2005, with permission from Forma 7.

help to prevent cancer, lower blood cholesterol, lower serum fat concentration-triglycerides in blood, and lower blood sugar in diabetics [48].

## 7.3.2 Genotypes

The ancestor of *Glycine max* (L.) Merr. was thought to be *Soja ussuriensis* Rgl. et Mack (*Glycine soja* Sieb. et Zucc.). According to geographical and climatic characteristics, *Glycine max* L. Merr. is grouped into six ssp. (*gracilis, indica, chinensis, corejensis, manchurica, slavonica*) with extensive morphological differentiations.

Cultivars are distinguished according to determinate and indeterminate growth, seed color, purpose of production (oil, protein), disease resistance, pests, drought, moisture, and length of growing stage, which lasts from 90 to 160 days. Numerous cultivars are divided into the categories early, medium early, and late. Very early cultivars (marked as 00) are more suitable for temperate conditions than cultivars (marked as 0). Both groups of cultivars are also marked with numbers from I to VI; groups III and IV occur too late for temperate climates. The cultivar needs to ripen before frost.

## 7.3.3 Morphology

Soy roots are well branched and. develop to plowing depth, for the most part. Although they can sometimes be found at a depth of 1.5 m 30 to 40 days after emergence, they will develop tuber-like formations due to infections from nitrogen bacteria (*Bradyrhizobium japonicum*). The number of tube-like formations varies (up to 90) and depends on the presence of bacteria.

Young plant stems are green with various color shades, up to light violet. The stem reaches a height between 0.2 and 1.0 m (with late cultivars and in favorable conditions, they may grow as tall as 2 m) and has 12 (15) internodes that grow between 3 and 13 cm in length. Between internodes there are nodes where plants sprout leaves and side shoots. The stem is branched and between 2 and 15 side shoots generally form. The first side shoots appear at a height between 1 and 20 cm and emerge from plumule nodes; thisoccurs mostly with early cultivars. The angle of side shoot growth varies from 50 to 70°, and branching depends on cultivar characteristics and growing conditions. Plants with high density branch only a little or not at all, and they are taller than lower-density plants. Plants growing in isolation develop a bush-like form.

Leaves are trifoliate with the exception of the first pair, which only grows one leaflet on the petiole. The petiole of the middle leaflet is the longest. Leaflets have different shapes (cordate, ovate, lanceolate) with pointed bracts; leaflet length is between 6 and 12 cm and width is between 3 and 9 cm.

The inflorescence of soy is *racemus*, forming 2 to 5 groups of flower (up to 20). Inflorescences are formed by composite leaves toward the top of the stem, reaching 3 to 15 cm above ground. Flowers can be short and tight or long and loose inflorescences. Flowers consist of five sepals, three petals, 10 anthers, and a pistil. Filaments of nine anthers are grown together, and the 10th anther (the upper one) is free. Determinate and indeterminate soy types are distinguished by inflorescence growth [47].

The flower petals may be white, pale violet, or pale yellow. Flowers open in the morning; the flowering stage depends on the earliness or lateness of the cultivar and can take between 15 and 80 days. Flowers have all of the characteristics of legumes and are 99.6% self-pollinated. Under normal production conditions, 20% of flowers fall off, but with temperatures under 10°C, high temperatures, or a lack of moisture, more than 90% of flowers can fall off.

Pods are 3 to 5 cm long with one 5 seeds and 2 to 3 seeds in most cases (Figure 7.2), and between 10 and 300 seeds can be expected per plant. Before maturing, green pods turn from light to dark yellow. The average plant grows 25 to 35 pods, but this number varies considerably. The opening of pods at full maturity is a varietal characteristic.

The seed consists of seed coat or testa (7 to 8% of joint seed weight) and embryo. The embryo consists of two cotyledons, plumula (embryonic shoot), and radicle (embryonic root). Testa is light yellow, brown, green, black, or a combination of colors. Cotyledons represent 90% of seed weight and are yellow or greenish. The part that connects a seed to a pod is called a funicle, which is a bit darker than a seed. Seeds are round or ovate; from the side, they are usually flattened with expressed funicle and grow to between 5.7 and 14 mm long. Absolute seed weight is 100 to 300 (40 to 450) g, and hectoliter weight is 85 kg hl$^{-1}$.

*Figure 7.2* Pods of soya bean.

## 7.3.4 Susceptibility to climate circumstances

Soy requires similar growth circumstances as corn and thrives where the sum of active temperatures in the growing stage amounts to at least 1700°C. Soy germinates at 8 to 10°C and emerges at 10 to 12°C. The optimal flowering temperature is 25 to 28°C (minimal is 17 to 18°C), pod formation temperature is 21 to 23°C (minimal is 13 to 14°C), and maturity temperature is 8 to 9°C. Soy normally matures at 16°C, butsome believe that the optimal maturing temperature ranges between 19 and 20°C.

Young emerged plants can survive spring frosts to –2.5°C without great damage; in early autumn, the crop is resistant to a temperature of –3°C. With temperatures that exceed 33°C and a lack of moisture in the flowering period, over 90% of flowers can fall off. Soy requires a great deal of moisture up to the flowering stage and afterward; during flowering, rain and excessive irrigation negatively influence pollination. Soy transpiration coefficients vary between 500 and 600 (390 to 750). In temperate conditions, precipitation is not a limiting production factor but distribution is. The most important factor is relative air moisture, which should be between 78 and 80%. Soy is a short-day plant. If produced in long-day circumstances, green weight growth increases and plants flower and mature later.

## 7.3.5 Cultivation according to organic guidelines

Soy cannot be grown as a monoculture. Suitable precrops are stable, manure-fertilized arable crops, stubble cereals, and oilseeds. Sunflowers are not included in alternative production crop rotation. Soy is a recommended precrop for most crops, though it is more difficult to optimize nitrogen fertilization with stubble crops because approximately 40 to 75 kg N ha$^{-1}$ remains in the soil with nodules or bacterial tubers. Soy is most successfully produced in fertile, light, deep, and slightly moist neutral soil; swampy soil is inappropriate, as are clay and sandy soils. Soy demands structured soil with a regulated water-air regime. In Slovenia, the most appropriate soil type is deep sandy-clay brown soil. Salty and acid soils are not recommended for soy; it grows well with a pH between 6.6 and 7.2.

Basic and precrop soil cultivation is the same as for early crops. Plowing should be done in autumn to the plowing depth. Roughly evening the soil also makes it easier to prepare the field with the presowing machine and an even sowing layer.

### 7.3.5.1  Fertilization

Between 1000 and 3500 kg seed ha$^{-1}$ can be expected. Nutrient uptake with 100 kg of seed included aboveground mass vary between 7.2 and 10.8 kg of nitrogen, 1.1 and 4.0 kg of phosphorus, and 3.0 and 6.2 kg of potassium. These amounts can be used to calculate the total of nutrients according to soil composition. Soy can also take 50 to 75% of nitrogen from the air through nitrogen bacteria, though this requires a suitable inoculation with relevant symbiotic nitrogen bacteria (*Bradyrhizobium japonicum*). Phosphorus and potassium fertilizers are plowed individually in autumn (one-third to one-half in spring) or in combination with 20 to 30 tons of stable ha$^{-1}$; during the first growth stages, the plants need available nitrogen in the soil (approximately one-third, from calculated needs). Animal manures and plant composts are more recommendable for fertilization and top dressing. For top dressing before the flowering stage, fertilizers for organic production can be used for plants not forming tubers. Because soy can use nutrients that are not easily accessible, yield is high even with poor fertilization but it leaves soil poor, which negatively influences the following yield.

Bacterial inoculation of seed is done with special preparations (nitragin, radicin, azotofiksin, and so forth) of the chosen soy-symbiotic bacteria. Seed is inoculated by mixing it with pulpy preparation directly before sowing. Infected moist seeds are protected from direct sun heat at sowing.

The sowing date is the same as for corn, or it follows immediately if corn is sown early. Late sowings in spring are also successful, but only with early cultivars. Stubble-crop sowing after early potatoes and barley is successful only with suitable irrigation. The sowing rate is between 80 and 130 kg seed ha$^{-1}$.

Sowing can be performed to depths of 3 to 5 cm with a corn-sowing machine. For home production, however, sowing may be done in heaps. If there is a danger of drought rolling should be performed immediately after sowing. Crusty soil is crushed before emergence; later on, crust is removed only between rows.

Weeds are removed with interrow hoeing, which preserves soil structure. Cultivation should take place to a depth of 10 cm when the first trifoliate leaf appears, and another shallow hoeing follows after 14 days. Later, hoeing is done when necessary until rows are joined.

Weed suppression can be avoided by sowing in weed-free soil. The use of herbicides in classical production is limited to rows, due to ecological damage. Hoeing should be done between rows. Irrigation takes place during the critical development stages before flowering, and also during pod formation and grain assimilation. Approximately 30 to 40 mm of water are used in one ration, and sprinkled if possible. As with other field crops, attention must be paid to soil pests (wireworms, chafer grubs, weevils, cutworms) and spider mites.

The soybean can be affected by some plant diseases. Mildews can be suppressed by somewhat effective sprays, while other diseases are closely connected to inappropriate crop rotation (e.g., white rot or *Sclerotinia sclerotiorum* appears in sunflower crop rotation, and root and stem rots appear with other crops).

### 7.3.6 Harvesting

Maturing is initiated by late drying of leaves, stems, and pods. The plant part becomes harder, the seed colors according to variety, and stems and pods turn brown. When 95% of pods obtain the significant color, approximately 10 days without rain are required to enable combining; seed can contain only 15% moisture. At the stage of over-maturity, the amount of losses from pods can exceed 35%. Therefore, we begin combining somewhat early, during uneven crop maturity, when pods are not yet completely ripe. Harvesting combines are adjusted by increasing the number of scythe movements, reducing the number of turns at the combine outlet, building a rubber curtain at the elevator entrance, adjusting the height of the reel, and surrounding it with sponge. All of these procedures result in yield loss. However, soy may be dried from small fields in heaves. Storage seed should be dried to contain less than 13% moisture and is stored in the same way as cereals.

## 7.4 Vigna: a few words about a widely spread genus

Vigna is a genus from the family *Fabaceae*, which have been described in the past as an Asian group of beans from genus *Phaseolus*. Vigna originated in Asia and Africa, but during the last two decades the interest in breeding and production programs has increased throughout the world, especially in the U.S. Important producers include China, India, Japan, Korea, Niger, Mexico, Cuba, Indonesia, Malaysia, New Zealand, and Australia. The crop requires tropical growth conditions and only a few species are suitable for production in temperate climates.

The *Vigna* sp. is one of the 12 most important grain legumes in the world, and its seeds contain between 19 and 24% proteins, depending on genotype. Some species, like *Vigna radiata* and *Vigna angularis*, are utilized for the production of edible sprouts; around 1 kg of seeds has the potential to produce 0.8 kg of sprouts. So-called "azuki" beans (azuki bean = adzuki = adsuki) are important for production, as well. The name, directly translated from Chinese, means "small bean." Azuki belongs to the Asian subgenus *Ceratotropis*, which includes 17 species, 7 of which are species cultivated with wild ancestors. Due to numerous morphological differences, systematists first submitted the Azukia species to the genus *Dolichos*, which was later reclassified into *Vigna*. Many synonyms appearing in folk and expert nomenclature cause problems and misunderstandings, because there is no clear overview of various systematics. The synonyms for important cultivated *Vigna* species or Adzuki beans, mungo, and others without spontaneously grown species are worth reviewing (Table 7.2).

The group of mungo beans (*Vigna radiata, V. mungo,* and *V. aconitifolia*) grown mainly in India features epigeal germination, with the visible primary leaf of the seedling. Azuki beans (like *V. angularis* and *V. umbel latta* from East Asia) feature hypogeal germination, when the first dicotyledonous leaves stay underground and the epicotyl part (the prolonged small stem with the first leaf) looks similar to the pea [49].

Identification for some *Vigna* species [49–53] is as follows:

- *Vigna unguiculata*: includes three ssp. *cylindrica* (7 to 12 cm long pods), ssp. *sesquipedalis* (30 to 90 cm long pods), and ssp. *unguiculata* (10 to 30 cm long pods). This plant is closely related to azuki and mungo beans: azuki beans grow 0.6 to 3 m high, while mungo beans are lower (0.3 to 0.9 m).
- *Vigna aconitifolia*: leaflets deeply lobed with 3 to 5 narrow lobes, pods glabrous, and stipules small to lanceolate.
- *Vigna mungo*: leaflets entire or very occasionally with 2 to 3 shallow broad lobes; plant and pods have hairy dull seeds and a short petiole on the primary leaf of the seedlings.
- *Vigna radiate*: leaflets like *V. mungo*, pods point sideways or down, covered with short hairs, seeds nearly or quite globular, brown to black, and hilum not concave.
- *Vigna angularis*: plants and pods glabrous or nearly so, seed smooth to shiny, a long petiole on primary leaf of the seedlings, pods constricted between seeds, and hilum not concave.
- *Vigna umbellata*: like *V. angularis*, but pods not constricted between seeds and hilum concave.

The inflorescence structure is similar to that of legumes. Seed color is not a reliable indicator for distinguishing species because it can vary with new varieties from chestnut brown, blue-black, black, green, ash grey, brown, grass green, and white to combination of colors; in production, red seeds prevail.

Yield varies from 300 to 3500 kg seed ha$^{-1}$ depending on species, varieties, and climate. Besides its economical function, production of this legume

*Table 7.2* Cultivated (C) and Cultivated-Wild (CW) Vigna Species [1–3]

| Actual Nomenclature/ Morphology | C, CW, 2n (group) | Synonyms | Common Name |
|---|---|---|---|
| *Vigna aconitifolia* (Jacq.) Maréchal | KD, 22 mungo | *Phaseolus aconitifolius* Jacq.; *P. palmatus* Forsk.; *Dolichos dissectus* Lam. | moth bean, mat bean |
| *Vigna angularis* (Wild.) Ohwi in Ohashi | CW, 22 azuki | *Phaseolus angularis* (Willd.) Wight; *Dolichos angularis* Willd.; *Azukia angularis* (Willd.) Ohwi | Azuki bean |
| *Vigna angularis* var. *angularis* | C, 22 | | |
| *Vigna glabrescens* M. M. & S. | CW, 44 | *Phaseolus glaber* Roxb.; *P. glabrescens* Steudel; *Vigna reflexopilosa* Hayata; *V. mungo* var. *glabra* (Roxb.) Baker; *V. radiata* var. *glabra* (Roxb.) Verdc. | tetraploid vigna |
| *Vigna mungo* (L.) Hepper | CW, 22 mungo | *Phaseolus mungo* L.; *P. Viridissimus* Ten.; *Azukia mungo* (L.) Masamune | blackgram, black mapte, kalai mash, mungo, urd, |
| *Vigna mungo* var. *mungo* | C, 22 mungo | *Phaseolus mungo* L. | urid, woolly, pyrol |
| *Vigna radiate* (L.) Wilczek | CW, 22 mungo | *Phaseolus radiatus* L.; *P. mungo sensu* F. B. I. non L.; *P. radiatus* var. *typicus* Matsum; *P. aureus* Roxb.; *Azukia radiata* (L.) Ohwi; *Rudua aurea* (Roxb.) Maekawa; *Vigna aureas* (Roxb.) Hepper; *V. mungo* Hepper, non (L.) | chickasaw pea, bundo, goldengram, greengram, moong, mung, mungbean |
| *Vigna radiate* var. *radiate* | C, 22 mungo | *Phaseolus radiatus* L.; *P. aureus* Roxb.; *Azukia radiata* (L.) Ohwi; *Rudua aurea* (Roxb.) Maekawa | mungo, Oregon pea, Yae-nari |
| *Vigna umbellata* (Thunb.) Ohwi & Ohashi | CW, 22 azuki | *Dolichos umbellatus* Thunb., *Phaseolus pubescens* Blume; *P. calcaratus* Roxb. *P. chrysanthus* Savi; *P. torosus* Roxb.; *P. ricciardianus* Tenora; *Vigna calcarata* (Roxb.) Kurz; *Azukia umbellata* (Thunb.) Ohwi | rice bean, climbing mountain bean |
| *Vigna umbellata* var. *umbellate* | C azuki | *Phaseolus calcaratus* var. *major* Prain; *P. calcaratus* var. *rumbaiya* Prain | Jerusalem pea, Mambi bean, red bean, oriental bean, Pegin bean, tzuruazuki, kanime |

*(continued)*

*Table 7.2* Cultivated (C) and Cultivated-Wild (CW) Vigna Species [1–3] (Continued)

| Actual Nomenclature/ Morphology | C, CW, 2n (group) | Synonyms | Common Name |
|---|---|---|---|
| *Vigna unguiculata* (L.) Walp. ssp. *cylindrica* (L.) Vedrc. | C biflora | *Dolichos biflorus* L.; *Dolichos catiang* L.; *D. catiang* Burm.; *Phaseolus cylindricus* L.; *Vigna catjang* (Burm. f.) Walp.; *V. cylindrica* (L.) Skeels | catjang, sow-pea, Catjangbohne-nem |
| *Vigna unguiculata* (L.) Walp. ssp. *sesquipedalis* (L.) Vedrc. | C unguiculata | *Dolichos sesquipedalis* L.; *Vigna sesquipedalis* (L.) Fruwirth; *V. sinensis* ssp. *sesquipedalis* (L.) Van Eselt | yard-long bean, pea bean, asparagus bean |
| *Vigna unguiculata* (L.) Walp. ssp. *unguiculata* | C unguiculata | *Dolichos sinensis* L.; *D. unguiculatus* L.; *Phaseolus unguiculatus* (L.) Piper; *Vigna sinensis* (L.) Savi & Hassk. | southern pea, cowpea, black-eyed pea, crowder pea |
| *Vigna subterranea* (L.) Vedrc. | C | *Glycine subterranea* L.; *Voandzeia subterranea* (L.) Thouars & DC. | bambarra groundnut, grund-bea, Congo goober, hog peanut |

provides the possibility of symbiotic nitrogen fixation with relevant strains from the genus *Bradyrhizobium*; provision with mycorrhiza from the genus *Glomus* in also possible. Nitrogen fixation amounted to between 1 and 99 kg N ha$^{-1}$ in tests.

Organic production from nonbean-growing regions is not well developed, but some Web sites feature organically grown products from the genus *Vigna*. For the species *V. unguiculata* ssp., similar growing conditions as the common bean (*P. vulgaris*) may be utilized. This species prefers nighttime temperatures of 18°C and daytime temperatures between 20 and 25°C; these temperatures are utilized for some species of vigna produced in greenhouses in the Netherlands. Two seeds of m$^{-2}$ are used for high plants, the same as for the climbing or runner bean; for bush types, same rules are applied as for the bush bean. Weed control is more critical for vigna than for many other dry, edible grain legumes, because the thermophilic vigna has a slow growth rate in the spring. Weeds can be controlled with cultivation after the second true leaf has developed on the seedlings, normally after about 3 weeks.

Growth conditions for vigna are closely related to the growth conditions of the soybean or bush bean. Azuki beans grow successfully in 30 to 36°C and can survive extreme drought.

# References

1. Langer, R.H.M. and Hill, G.D., *Agricultural Plants*, Cambridge University Press, Cambridge, 1993, 217.
2. Frame, J., Chalton, J.F.L., and Laidlaw, A.S., *Temperate Forage Legumes*, CAN International, Wallingford, New York, 1997.
3. Yadaya, N.D., *Arid Legumes*, Agrobotanical Publishers, New Delhi, 1999.
4. Kumarasinghe, K.S., Danso, S.K.A., and Zapata, F., Field evaluation of N-2 fixation and N-partitioning in climbing bean (*Phaseolus vulgaris* L.) using N-15, *Biol. Fert. Soils*, 13, 142, 1992.
5. Hermann, G. and Plakolm, G., *Organic Farming (Husbandry for Practice)*, Verlagsunion Agrar, Vienna, 1993, 10.
6. Bavec, F. et al., Competitive ability of maize in mixture with climbing bean in organic farming, in *Researching Sustainable Systems*, presented at First Scientific Conference of the International Society of Organic Agriculture Research (ISOFAR), Adelaide, Sept. 21–23, 2005, Köpke, U. et al., Eds., IFOAM, Adelaide, Australia, 2005, 502.
7. Anil, L., Park, J., and Phipps, R.H., The potential of forage-maize intercrops in ruminant nutrition, *Anim. Feed Sci.Technol.*, 86, 157, 2000.
8. Deshpande, S.S., Food legumes in human nutrition: A personal perspective, *Crit. Rev. Food Sci. Nutr.*, 32, 333, 1992.
9. Iqbal, A. et al., Nutritional quality of important food legumes, *Food Chem.*, 97(2), 331, 2005.
10. Burn, E.E. and Huffaman, V.L., Food quality and peanut products, in *Science and Technology of Groundnut: Biology, Production, Processing, and Utilization*, Desai, B.B., Kotecha, P.M., and Salunkhe, D.K., Eds., Naya Prokash, Calcutta, 1999.
11. Bavec, F., *Nekatere Zapostavljene in/ali Nove Poljščine (Some of Disregarded and/or New Field Crops)*, Univerza v Mariboru, Fakulteta za kmetijstvo, Maribor, 2000.
12. Amjad, I., Khalil, I.A., and Shah, H., Nutritional yield and amino acid profile of rice protein as influenced by nitrogen fertilizer, *Sarhad J. Agric.*, 19, 127, 2003.
13. Farzana, W. and Khalil, I.A., Protein quality of tropical food legumes, *J. Sci. Technol.*, 23, 13, 1999.
14. Bavec, F., *Cicer arietinum* (L.), in *Nekatere Zapostavljene in/ali Nove Poljščine (Some of Disregarded and/or New Field Crops)*, Univerza v Mariboru, Fakulteta za kmetijstvo, Maribor, 2000, 28.
15. Langer, R.H.M. and Hill, G.D., *Agricultural Plants*, Cambridge University Press, Cambridge, 1993, 217.
16. Nayyar, H. et al., Chilling effects during seed filling on accumulation of seed reserves and yield of chickpea, *J. Sci. Food Agric.*, 85 (11), 1925, 2005.
17. Clarke, H.J. and Siddique, K.H.M., Response of chickpea genotypes to low temperature stress during reproductive development, *Field Crops Res.*, 90 (2–3), 323, 2004.
18. Anwar, M.R., McKenzie, B.A., and Hill, G.D., The effect of irrigation and sowing date on crop yield and yield components of Kabuli chickpea (*Cicer arietinum* L.) in a cool-temperate subhumid climate, *J. Agric. Sci.*, 141 (3–4), 259, 2003.

19. Pramanik, K. and Singh, R.K., Effect of levels and mode of phosphorus and biofertilizers on chickpea (*Cicer arietinum*) under dryland condition, *Indian J. Agron.*, 48 (4), 294,

20. Rudresh, D.L., Shivaprakash, M.K., and Prasad, R.D., Effect of combined application of *Rhizobium*, phosphate solubilizing bacterium and *Trichoderma* spp. on growth, nutrient uptake and yield of chickpea (*Cicer arietinum* L.), *Appl. Soil Ecol.*, 28 (2), 139, 2005.

21. Saini, V.K., Bhandari, S.C., and Tarafdar, J.C., Comparison of crop yield, soil microbial C, N and P, N-fixation, nodulation and mycorrhizal infection in inoculated and non-inoculated sorghum and chickpea crops, *Field Crops Res.*, 89 (1), 39, 2004.

22. Nautiyal, N. and Chatterjee, C., Molybdenum stress-induced changes in growth and yield of chickpea, *J. Plant. Nutr.*, 27 (1), 173, 2004.

23. Nautiyal, N., Singh, S., and Chatterjee, C., Seed reserves of chickpea in relation to molybdenum supply, *J. Sci. Food Agric.*, 85 (5), 860, 2005.

24. Mohammadi, G. et al., Critical period of weed interference in chickpea, *Weed Res.*, 45 (1), 57, 2005.

25. Chopra, N., Chopra, N.K., and Singh, H.P., Loss in seed yield and quality due to weed stress in chickpea (*Cicer arietinum*), *Indian J. Agric. Sci.*, 73 (6), 350, 2003.

26. Mauromicale, G. et al., Root nodulation and nitrogen accumulation and partitioning in legume crops as affected by soil solarization, *Plant and Soil*, 271 (1–2), 275, 2005.

27. Iqbal, A. et al., Nutritional quality of important food legumes, *Food Chem.*, 97(2), 331, 2005.

28. Rehman, Z. and Shah, W.H., Thermal heat processing effects on antinutrients, protein and starch digestibility of food legumes, *Food Chem.*, 91 (2), 327, 2005.

29. Abdel-Rahman, A.H.Y., Effect of cooking on tryptophan, basic amino acids, protein solubility and retention of some vitamins in two varieties of chick pea, *Food Chem.*, 11 (2), 139, 1983.

30. http://www.FAO.ORG. (accessed 2002).

31. Samuel, E.J.J., Duraira, K.S.P., and Mohan, S., TLC for the detection of aflatoxin in groundnut (*Arachis hypogaea* L.) kernels, *Asian J. Chem.*, 14 (2), 874, 2002.

32. Bavec, F., *Arachis hypogaea* (L.), in *Nekatere Zapostavljene in/ali Nove Poljščine (Some of Disregarded and/or New Field Crops)*, Univerza v Mariboru, Fakulteta za kmetijstvo, Maribor, 2000, 162.

33. Desai, B.B., Kotecha, P.M., and Salunkhe, D.K., *Science and Technology of Groundnut: Biology, Production, Processing, and Utilization*, Naya Prokash, Calcutta, 1999.

34. Mallikarjuna Swamy, B.P. et al., Phenotypic variation for agronomic characteristics in a groundnut core collection for Asia, *Field Crops Res.*, 84 (3), 359, 2003.

35. Knauft, D.A. and Gorbet, D.W., Genetic diversity among peanut cultivars, *Crop Sci.*, 29, 1417, 1989.

36. Reddy, M.S., Floyd, C.N., and Willey, R.W., Groundnut in intercropping systems, in *Proc. Int. Intercropping Workshop*, Gibbons, R.W., Ed., ICRISAT, Andhra Pradesh, India, 1980, 133.

37. Willey, R.W. and Reddy, M.S., A field technique for separating above- and below-ground interactions in intercropping, an experiment with pearl millet-groundnut, *Exp. Agric.*, 17 (3), 257, 1981.
38. Ofori, F. and Stern, W.R., Cereal-legume intercropping systems, *Adv. Agron.*, 41, 41, 1987.
39. Ghosh, P.K., Growth, yield, competition and economics of groundnut/cereal fodder intercropping systems in the semi-arid tropics of India, *Field Crops Res.*, 88 (2–3), 227, 2004.
40. Rao, S.S. and Shaktawat, M.S., Effect of organic manure, phosphorus and gypsum on groundnut (*Arachis hypogaea*) production under rainfed condition, *Indian J. Agron.*, 47 (2), 234, 2002.
41. Elsheik, E.A.E. and Mohamedzein, E.M.M., Effects of biological, organic and chemical fertilization on yield, hydratation coefficient, cook ability and mineral composition of groundnut seeds, *Food Chem.*, 3 (2), 253, 1998.
42. Vasilevski, G., Nout — *Arachis hypogaea* (L.), in *Posebno Ratarstvo*, 1. Part, Jevti, S., Ed., Nauna knjiga Beograd, Beograd, 1986, 388.
43. Bankole, S.A., Ogunsanwo, B.M. and Eseigbe, D.A., Aflatoxins in Nigerian dry-roasted groundnuts, *Food Chem.*, 89 (4), 503, 2005.
44. IARC, *Some Naturally Occurring Substances: Food Items and Constituents, Heterocyclic Amines, and Mycotoxins: IARC Monographs on Evaluation of Carcinogenic Risk to Humans*, Vol. 56, International Agency for Research on Cancer, Paris, France, 1993.
45. Henry, S. et al., Aflatoxins, WHO, *Food Additive Series*, 40, 361, 1998.
46. Oettle, A.G., Cancer in Africa, especially in regions south of the Sahara, *J. Nat. Cancer Inst.*, 33, 383, 1964.
47. Bavec, F., Soyabean, in *Nekatere Zapostavljene in/ali Nove Poljščine (Some of Disregarded and/or New Field Crops)*, Univerza v Mariboru, Fakulteta za kmetijstvo, Maribor, 2000.
48. Pokorn, D., Soja osvaja svet (The soyabean to win a world), *Forma 7*, Ljubljana, 1995, 11.
49. Bavec, F., Vigna, in *Nekatere Zapostavljene in/ali Nove Poljščine (Some of Disregarded and/or New Field Crops)*, Univerza v Mariboru, Fakulteta za kmetijstvo, Maribor, 2000.
50. Lumpkin, T.A. and McClary, D.C., *Azuki Bean: Botany, Production and Uses*, CAB International, Wallingford, 1994.
51. http://www.bacnr.it/Beanref/sp_vgrin.htm (accessed October 1994).
52. Jain, H.K. and Mehra, K.L., Evolution, adaptation, relationships, and uses of the species of *Vigna* cultivated in India, in *Advances in Legume Science*, Summerfield, R.J. and Bunting, A.H.J., Eds., Indian Agriculture Research Institute, New Delhi, 1980, 659.
53. Tomooka, N. et al., *Vigna riukiuensis*, in *Mungbean and the Genetic Resources of the Subgenus Ceratotropis, in Ecological Studies on Tropical Food Legumes in Relation to Adaptation to Cropping Systems in Thailand*, Tropical Agriculture Research Center, Tsukuba, Japan, 1991, 14.

# chapter eight

# Traditional and new kinds of food from alternative crops

## 8.1 Homemade food from buckwheat

Utilization of buckwheat as a food depends on varying traditions in different regions throughout the world [1]. New trends of buckwheat utilization are based on tradition, culinary art, and growing knowledge about functional and healthy food, especially if it is organically produced. Diverse cooking methods for preparing food from buckwheat are used, such as boiling, baking, frying, and steaming. However, the main specialties made from buckwheat are bread, different kinds of noodles, boiled groats (kasha), pancakes and other cakes, polenta, soba-gaki or ûganci, buckwheat roles, buckwheat dumplings, popped buckwheat, sprouts, tea, and so forth. The groats fraction of the mill process is not suitable for making food based on buckwheat flour. Most buckwheat food can be eaten with vegetables, meat, or seasoning, embodying different traditions and tastes.

### 8.1.1 Buckwheat bread

#### 8.1.1.1 Standard recipe

Making bread solely from buckwheat flour is not possible, because buckwheat flour has no proteins such as gluten and other alcohol-soluble proteins, which form the soft bread center. Instead, buckwheat bread can be made with a mixture of approximately 30 to 35% organic buckwheat flour and 65 to 70% organic wheat flour with high protein quality. Use 0.5 l of buckwheat flour scalded with 0.5 l of boiling salted water, then knead. Separately, dilute 25 g of fresh yeast, a teaspoon of sugar, and a teaspoon of wheat flour and lukewarm milk or water. When the buckwheat dough is lukewarm, add 0.75 to 1.0 l of wheat flour and the risen yeast; mix dough until it is sticky and add water during kneading if the dough is too dry. After approximately 30 minutes, when the dough has risen at room temperature, put the dough into a smeared baking pan (without kneading). After 10 to 15 minutes, when the dough has risen again, the bread is ready for baking. Bake at 220°C for about an hour.

*8.1.1.1.1  Buckwheat bread with walnuts or large seeds.* Make everything according to the standard recipe. Before kneading the dough made from mixed wheat flour, yeast, and separately kneaded buckwheat dough, add some walnuts or large seeds, such as oil pumpkin.

*8.1.1.1.2  Variegated (spiral color) bread.* You can prepare buckwheat dough according to the standard recipe, using white wheat dough. When both kinds of dough have risen for the first time, prepare separate layers. Put the layer of buckwheat dough on top of the layer of wheat dough and roll them together. After the rolled dough has risen for 15 minutes, bake it like any other bread. Slices of finished baked bread should have bicolor spiral stripes.

Different kinds of bread are made from cereal mixtures such as spelt wheat and buckwheat. Spelt rye and buckwheat or common wheat, rye or common wheat, and maize and buckwheat can contain less than 30% of buckwheat flour. You should make buckwheat dough with boiling water, as previously described. A small portion of the buckwheat flour in the mixture does not need to be kneaded separately. A small part (up to 25%) of mixed bread can also be made from boiling potatoes or flour from mashed potatoes.

## 8.1.2  Dishes made from buckwheat groats

Many traditional dishes are based on buckwheat groats (husked buckwheat seeds, kasha), and many modern dishes can be made by replacing rice with buckwheat. Use buckwheat groats for inventive dishes with just a little imagination and a good sense of culinary art. Groats cook easily within 10 to 15 minutes; special care should be taken not to overcook them. If you serve dry groats, keep them covered and warm so they remain soft. Many different foods can be made or enhanced by groats, including vegetable and meat dishes, salads, filled chicken, duck or turkey, baked buckwheat groats, and Italian risotto. Buckwheat groats are often used in soups like cooked groats in beef broth, vegetarian buckwheat soups, or soups in combination with vegetables and meats.

### 8.1.2.1  Buckwheat kasha

Boiled groats as a side dish are often served with meat sauces, wine cream, sour cream, vegetable sauce, or herb sauce. To improve the taste, put sliced onion into the pan with a teaspoon of warm oil. When the onion is yellow, add 0.25 l of washed groats and 0.4 l of cold water, add some milled black pepper and salt, and cook just enough to retain a slightly firm texture. Remove from heat, cover, and serve when the groats are soft and dry.

### 8.1.2.2   Buckwheat dumplings

Put 0.25 l of well-washed groats in 3 to 4 teaspoons of hot oil and roast until soaked. Pour boiling water over groats, cover, and steam. Lightly mix 50 g of butter, three egg yolks, and cold groats; stir  and mix in whipped egg white. Form dumplings and cook them in salted boiling water. If a test dumpling falls apart, add bread crumbs. Serve with sauce, as mentioned previously [2].

### 8.1.2.3   Baked buckwheat kasha

Cook 0.5 l of groats for 20 minutes in salted 1.5 l mixture of milk and water. After approximately 10 hours, add 50 to 100 g ground and cut large pieces of walnuts, two eggs, and a small teaspoon of sugar. Grease a baking pan and pour a 1.5 to 2 cm thick layer of groats mixture onto it. Grate a layer of apples over another layer of groats and cover the top with cream made from egg and sour cream. Bake at 150°C for 40 to 50 minutes.

### 8.1.2.4   Salad from buckwheat groats

Cool-cooked groats, as mentioned previously, are suitable for mixtures with different vegetables. For an excellent taste, dressing made with pumpkin seed oil is recommended, as are vinegar and salt.

### 8.1.2.5   Chicken, duck, or turkey filled with buckwheat groats

Wash the groats and soak them in cold water for 20 minutes. Cook the groats for 15 minutes in salt water until they are somewhat firm; leave them covered for a few hours. Add sliced onion to hot grease until the mixture is slightly yellow in color. Add hash or grind poultry, pork, or other meat. Pork meat may be smoked, but all meat should be soft. For vegetarian dishes, substitute preferred vegetables for meat. Add cooked, dry groats and mix the stuffing. According to your taste preferences, add some garlic, parsley, or fruit (such as apple slices or soaked and hashed dry plums) to the stuffing. Spice meat with salt and appropriate spice mixture, fill with the groats mixture, and close with wooden pin. Put the stuffed meat in the pan, smear the outside with salt and fat, and add a whole apple, onions, garlic, and any other spices suitable for baking with poultry. Bake as you would bake poultry, but during the baking process, smear the meat with drippings  from bottom of the pan. If the meat is still too dry, add some water; at the end of the cooking time, you can also add a teaspoon of beer. The baking time required will be longer than poultry without stuffing.

### 8.1.2.6   Soups

Groats are a very suitable component for vegetable soups or soups mixed with smoked meats. Soups like porridge are traditional in some European countries.

### 8.1.2.7   Noodles

Various kinds of noodles (soba kiri, soba noodles, and so forth) are used in Asian cooking. Buckwheat noodles are prepared by cutting thinly spread dough made from various combinations of buckwheat flour and water without any additives, or they are prepared with binding additives such as wheat flour, hen eggs, and so on. Noodles are classified into three types: raw, dried, and instant. Numerous traditional and much entangled food processing, recipes, and dishes are described [3, 1]. For cold dishes, cut noodles are cooked in boiling water, removed from water after soaking, washed in tap water, and than served with other cold foods. Hot noodles are dished up with toppings (such as soy sauce) from hot sauce-based soups.

### 8.1.2.8   Homemade noodles

If you cannot buy organic noodles, it is possible to make them at home. The buckwheat flour must be fresh, without milled groat fractions. Mix 0.3 l of organic buckwheat flour, 0.3 l of organic white wheat flour, salt, and two eggs, and add warm water as necessary. The dough must be well kneaded and well rolled in order to form thin layers of the dough on a wooden cutting board; cut the stacked disc-shaped layers well. Cook in boiling water and serve as described earlier. Keep in mind that in Japan, noodles have excellent palatability and many people are fastidious about their quality.

### 8.1.2.9   Slovenian žganci, Italian polenta, and Japanese soba-gaki methods

These foods are based on buckwheat flour cooked in boiling water. The flour must be of the fresh, white, nonrobust, sandy-looking fraction.

For buckwheat ûganci, put 0.5 kg of buckwheat flour into 1.5 l of salted boiling water and cook covered for 6 minutes. Then make a hole in the flour with a ladle or mix it around, so that the upper layer becomes the bottom one. After 6 minutes, pour some of the water out and mix in žganci. If you want to crumble the žganci, add 25 g of hot fat and leave it covered for a few minutes. Than crumble the žganci into a warm dish and serve it larded with cracklings and mushroom or clear meat soup. If you cut the žganci with the spoon from the ladle, add some of the water that was poured away, so they remain soft.

### 8.1.2.10   Cooked buckwheat rolls (with cottage cheese or walnuts)

Scald 0.5 l of buckwheat flour (if you wish to substitute half of the amount with wheat flour, prepare scaled buckwheat flour first and then add the wheat flour). Form into smooth but firm dough. Roll the cool dough on a wooden table until the dough layer is only a few millimeters thick. Add mixed cream made from cottage cheese (0.5 to 0.7 kg), 0.2 l of cream, 2 eggs, and 3 to 4 teaspoons of bread crumbs to the flattened dough. Roll it and wrap it in a kitchen cloth, bind it, and cook it slowly in boiling water for 30 to 45 minutes. Serve as a cake or garnish with sauce made from game meat or something similar.

You can replace the mixed cream with a popular filling made from a mixture of 0.3 l of ground walnuts, 4 teaspoons of cream, 1 egg, and 1 to 2 teaspoons of bread crumbs. Sugar, vanilla, cinnamon, or buckwheat honey can be added to the mixtures of both fillings if the roll is served as a cake. Top the cake with caster sugar or bread crumbs, fried in butter or oil.

### 8.1.2.11  Buckwheat pancakes

These pancakes of different thickness, taste, and ingredients (such as Indian pharpar with sesame) are one of the most common food types in many countries. You can make pancakes by using buckwheat flour or a mixture of wheat and buckwheat flour, milk, sugar, and salt. For a foamy consistency, you should use a mixture of both flours and add baking powder and whipped egg white. The procedure for baking is the same as for other pancakes. Use your imagination when adding ingredients, creams, or marmalades. The authors' favorite pancakes are rolled and filled with a mixture of cream, cottage cheese, buckwheat honey, and vanilla. Another filling is made of ground walnuts, cream, hot milk, and sugar or honey. Put the rolled pancakes in a pan and bake them at a high temperature; serve warm.

### 8.1.2.12  Buckwheat cake

The following recipe is for a buckwheat roll with walnuts, also called a Slovenian buckwheat "Potica" roll.

Scald 0.5 l of buckwheat flour with 0.5 l of salted boiling water. Add fermented yeast made from 30 g of leaven and some warm milk, along with a small teaspoon of sugar to the warm dough. Add 0.75 l of white wheat flour, then knead. When the dough has risen, flatten it with a rolling pin. For the filling, mix together 5 spoons of cream, sugar, or buckwheat honey, 0.5 l of ground walnuts, a few teaspoons of bread crumbs, and an egg. Add the filling to a thin layer of dough and roll it; place it in a greased baking pan, wait a few minutes for the roll to rise again, and bake it at 220°C. Garnish with caster sugar and serve [adapted by 2].

## 8.2   Homemade food from amaranths

### 8.2.1   Popped seeds

Traditionally amaranths were — and still are today — popped by contact with direct heat on a warm surface. Previously cleaned seeds are laid on a plate heated to around 180 to 200°C. The popped seeds should be removed from the hot surface after about 20 to 30 seconds; otherwise, the seeds will begin to burn. Popped seeds are eaten as a snack or light meal with milk or yogurt, and they are also used in cakes, home-made bread, granola bars with honey, seeds, nuts, muesli, and so forth.

## 8.2.2   Bread

Bread making is most successful when using a combination of 80% wheat flour and just 20% additional amaranth flour (or different milling fractions) without decline of quality, because amaranth seeds do not contain gluten. When making pan bread, amaranth can be added raw, cooked, popped, or as whole milled seed (or it can be milled after dehulling). Row seeds can soak a day before preparing the bread. Wholemeal bread should not be called wholemeal product if the amaranth is dehulled prior to its addition.

When preparing dough and baking bread, the standard recipes for home-made bread made from wheat flour are recommended. Instead of 100% wheat flour, you may substitute the mixture discussed here.

## 8.2.3   Pasta and noodle products

To make pasta and noodles, the products shoulld contain no less than 90% white wheat flour, not more than 10% additional amaranth flour (or cooked amaranth seeds), eggs, and salt water; they should be made by following the standard procedure for homemade noodles. The dough should be very tough after kneading, in order to be rolled into thin pieces. Fresh noodles should cook in salted boilling water for approximately 10 minutes; dry noodles require a little more time.

# 8.3   Homemade food from millet

## 8.3.1   Millet kasha with ground meat

For 4 servings, you will need 500 g of ground meat, 370 g of millet, 75 g of oil, 1 small onion, 150 g of grated carrots, kohlrabi, or celery, 2 peeled and cubed tomatoes, 2 cloves of garlic, 2 eggs, parsley, salt, pepper, and grated cheese.

Fry the onions and meat together in the oil. When the meat becomes softer, add garlic, carrot, kohlrabi, or celery, tomatoes, salt, and pepper. Stew for 10 minutes and then add washed millet and 0.7 l of water. Cook over moderate heat until the millet softens. Pour the resulting paste into a greased-baking pan over scrambled eggs, sprinkle with grated cheese, and bake for 10 minutes. Serve with salad.

# References

1. Kreft, I. et al., *Ethnobotany of Buckwheat*, Jinsol Publishing Co., Seoul, 2003.
2. Kalinšek, F., *Kuharica s. Felicite Kalinšek*, Cankarjeva založba, Ljubljana, 2004.
3. Udesky J. and Sturtleff, W., *The Book of Soba*, Harper & Row, New York, 1995.

# Index

## A

Accreditation, 8
Additives, 1, 17
Alkaloids, 150–152, 154, 157
Allelophaty, 70, 182
Allergies, 76
Alternaria
  *brassicae*, 139
  *carthami*, 144
*Amaranthus* sp., *see also* Grain amaranths
  *caudatus*, 32, 88–89, 90-93, 95–97
  *cruentus*, 31, 88–89, 90, 92-96
  *hybridus*, 92, 94
  *hypochondriacus*, 31, 88–89, 90, 92-96, 98
  *mantegazzianus*, 93
  *paniculatus*, 89, 96
Amino acids, 43-44, 49, 57, 74, 78, 85–86, 94,
          96–97, 116, 118, 120, 123, 124, 136,
          177, 202, 207, 212–213
Animal welfare, 2, 3, 4, 12, 15-16
*Anthalia* ssp., 149
Antioxidants, 66, 74-75, 100, 109, 136, 194
*Aphanomyces cochlioides*, 149
*Arachis* sp., 201
*Arachis hypogaea*, 31, 207–208, *see also*
          Groundnut
*Aspergilus flavus*, 211
*Avena fatua*, 51

## B

Barnyard grass, 31, *see also* Barnyard millet
Barnyard millet, 109, 123
*Bifidobacterium lactic*, 75
Biodiversity, 1, 2, 8, 17–21, 28–29, 33, 83, 88,
          149
Biodinamyc agriculture, 5

Biological diversity, 2, 4, *see also* Biodiversity
Brassica
  sp.
  *alba*, 30, 146–147
  *hirta*, 147
  *juncea*, 147, 149
  *napus*, 30, 32
  *nigra*, 147, 149
*Bradyrhizobium*, 210
Broomcorn millet, 32, *see also* Proso millet
Buckwheat, 30, 65–76
  allelophatic potential, 70
  allergies, 76
  amino acids, 65, 74
  autoclaving process, 77
  beehives, 65, 71
  botany, 66
  breads, 78
  carbohydrates, 73
  centuries, 65
  chemical components, 68
  climatic conditions, 68, 72
  crop rotation, 70, 72
  cultivars, 67, 70–72
  day length, 69, 71
  ecology, 68
  emergence, 69
  exporters, 78
  fats, 68, 73
  fertilization, 66, 69–70, 72
  fibers, 73
  flavanoids, 73, 75
  flours, 73, 74, 76–77
  foods, 66, 68, 74–77
  full-season, 68, 71–72
  genotypes, 65–66, 69, 71–72, 76
  grinding stone, 76
  groats, 73, 75–77
  growth, 66-70

harvest, 66, 69–70, 72
health, 73–74
homemade food, 225–229
honey, 65–66
hulled grains, 76
humus, 70-72
inflorescence, 66
interrow space, 71
kasha, 76–77
land race populations, 66, 71
milling 73, 76
mineral content, 74-75
morphology, 66
nutrients, 72
nutritional value, 65, 73–74
oils, 73
organic cultivation, 70
organic production, 68
photoperiodic reactions, 71
plant population, 69, 69–71
processing, 76-78
produced organically, 71–72
producers, 68
production, 65, 68, 70–72, 76, 78
products, 70, 73, 76, 77-78
proteins, 68, 73–74, 76
soba noodles, 75–76, 228
sowing, 66, 69–72
storage, 72
taxonomy, 66
technological value, 76
thermal treatment, 76
transpiration coefficient, 69
uptake efficiency, 72
vegetation period, 72
vitamins, 65, 74–75
yields, 65, 68–72
weeds, 70–71

# C

CAP (Common agriculture policy), 20
*Camelina,*
   sp.
   *alyssum, 137*
   *microcarpa, 137*
   *rumelica, 137*
   *sativa,* 31, 136–137, 139, *see also* Camelina
Camelina, 31
   *Brasiaceae,* 137
   amino acids, 136
   botany, 137
   cholesterol content, 136
   climates, 138

cold-pressed oils, 139
cultivars, 138-139
genotypes, 136–137
health value, 136
maturity, 139
morphology, 137
nutritional value, 136
oils, 136–139
products, 139
production, 136, 138–139
storage, 139
traditional medicine, 137
utilization, 136,139
weeds, 137
Chenopodium
   *quinoa,* 31–32, 78–79, *see also* Quinoa
   *berlandierri, 79*
   *pallidicaule, 79*
   *nuttalliae, 79*
*Cannabis sativa,* 30, 32, 163, 166, 169–170, *see*
          *also* Industrial and edible–seed
          hemp
Carthamus
   sp.
   *lanatus,* 140
   *oxyacantha,* 140
   *tinctorius,* 31–32, 139–140, *see also*
         Safflower
Certification, 7, 8, 10, 11
*Cicer* sp., 201, 203
*Cicer arietinum,* 30, 202, *see also* Chickpea
*Cichorium intybus,* 32
Ceutorhynchus
   sp.
   *assimilis,* 149
   *maculaalba,* 156
*Chenopodium,* 30–32
   sp.
   *album,* 79
   *berlandierri,* 79
   *pallidicaule,* 79
   *quinoa,* 31–32, 78–79, *see also* Quinoa
Chicory, 32
Chickpea, 30
   amino acids, 207
   botany, 203
   climates, 204, 206
   cultivars, 203, 205
   cultivation, 203, 205
   development, 203–204
   digestibility, 207
   fertilizers, 205
   germination, 203, 206
   growth, 203–204, 2006
   maturity, 203, 206

molybdenum, 206
nutrient uptake, 206
proteins, 202, 204, 206, 207
regions, 2003
*Rhizobium*, 2003, 205
soil-borne pathogens, 206
soils, 205–207
transpiration coefficient, 204
water stress, 204
weeds, 206–207
yields, 204–207
*Claviceps purpurea*, 54
*Crambe abyssinica*, 32
Codex Alimentarious, 2, 7, 13, 25
Conventional farming, 1–2, 19–20
*Coriandrum sativum*, 32
Crop rotation, 2, 5, 17, 22–23, 26, 28–30, *see also* each chapter of crops
Conservation, 6, 20–21, 25, 27–28, 33
*Cucurbita pepo*, 30, 127, *see also* Oil (seed) pumpkins
*Cuphea* spp., 32

# D

Day length, 69, 71, 142, 148
*Dasyneura papaveris*, 156
*Digitaria*, 31, 109, 121
   sp.
   *exilis*, 31, 122, *see also* White fonio
   *iburua*, 122
Digestibility, 88, 121, 202, 207, 212
*Dimorphotheca pluvialis*, 32

# E

Echinochloa
   *colona*, 31, 32, 123
   *crus-galli*, 31, 32, 109, *see also* Barnyard millet
Einkorn, 30
   amino acids, 49
   botany, 49
   chemical composition, 49
   climatic circumstances, 51
   cultivation practice, 49
   fertilizers, 49
   fibers, 49–50
   flours, 50
   foods, 50
   gluten, 50
   growing conditions, 49
   hulled seeds, 49
   hulled wheat, 50

maturity, 50
processing, 49
production, 48
proteins, 49–50
salads, 48
seeding rates, 49
sedimentation, 50
yields, 49–50
*Eleusine coracana*, 31–32, 109, 119, *see also* Finger millet
*Elytrigia intermedia*, 31, *see also* Intermediate wheatgrass
Emmer, 31
   botany, 50
   history, 50
   nutritional compounds, 51
   organic farming, 51
   origins, 50
   production, 50–51
   soils, 50
   utilization, 51
   water deficiency, 50
   yields, 51
Environmental impacts, 20, *see also* Organic agriculture
*Eregrostis tef*, 109, *see also* Tef
*Erwinia* spp., 156
*Euphorbia* spp., 32

# F

FAO (Food and Agriculture Organization), 7, 78, 85, 97, 143
*Fabaceae*, 201, 203, 217
Fagopyrum
   *esculentum*, 30, 65–66, *see also* Buckwheat
   *tataricum*, 66
   *cymosum*, 66
Farro, 38
Fenugreek, 32, 201
Finger millet, 31–32
   amino acids, 120
   biology, 120
   foods, 119–120
   food processing, 121
   genotypes, 120
   germination, 121
   growth, 120
   nutritional value, 120
   production, 119-120
   traditionally used, 121
   utilization, 120
Flax, 30–32
   amino acids, 185

ancestor, 178
botany, 179
breads, 184–185
cakes, 184–185
climates, 178, 181
cultivars, 179, 180, 183
cultivation, 182–183
development stages, 180
edible seeds, 178
fibers, 179, 184–185
fertilizers, 182–183
growing period, 181–182
growth conditions, 181, 183
growth stages, 176, 179–180
harvest, 180,184
maturity, 181, 184
medicine, 185–186
morphology, 179
nitrogen use, 182
oils, 181, 185
photosynthesis, 181
polluted regions, 185
production, 178–179, 182–185
remediation, 178
seed rates, 183
sowing, 179, 182–184
textile, 178
utilization, 184
weeds, 182, 184
yields, 179, 182–183
Flour mixture, 46
Fonio, 31, *see also* White fonio
Food and Agriculture Organization, see FAO

Food safety, 1, 3, 7–8, 12
Foxtail, foxtail millet, 30, 111
   biological characteristics, 114
   cultural practice, 116
   cultivars, 115
   growing conditions, 114
   nutrient content, 115
   production, 114, 116
   products, 116
   proteins, 114, 116
   utilization,116
   yields, 115–116
*Fusarium* spp., 58, 71, 144, 156

**G**

*Gaeumannomyces* sp., 70
Garden poppy, 30
   alkaloids, 150–152, 154, 157
   cultivars, 151–157

consumption, 155, 157
gene center, 150
growth,  153–154, 154
growth stages, 153–154
harvest, 156–157
medicines, 150–151
morphology, 151
nutritional composition, 156
oils, 156-157
opiates, 150–154, 156-157
organic cultural practice, 154
pharmaceutical industry, 150
precrops, 154
production, 153–154, 156
products, 151, 157
spring sown, 155
urine, 157
uses, 155–156
vegetation period, 155–156
winter, 154–155
weeds, 150, 154–156
yield, 156
Genetically modified organisms, see GMO
*Glycine* sp., 201
*Glycine max*, 211, 213, *see also* Soybean
Gluten-free, 45, 51, 98, 116
GMO, 1, 4, 8, 12, 14–15, 29
Grain amaranths, 31
   amino acids, 94, 96
   blood-anemia, 98
   botany, 89
   carbohydrates, 94
   civilizations, 89
   cultivation, 88–89, 92
   diets, 98
   direct seeding, 93
   ecology, 90
   foods, 88–89, 94–95, 97–98
   food processing, 95, 97
   glutelins, 96
   gluten-free, 88–98
   growth, 91–93
   harvest, 92–94
   harvest index, 92
   health, 98
   human nutrition, 88–89, 98
   illumination, 91
   maturity, 92,94
   minerals, 94
   nutrients, 92–94
   nutritional value, 94
   oils, 95, 98
   organic cultural practice, 92
   organic fertilizers, 93
   organic foods, 95, 97

percentage of live seed emergence (PLSE), 91, 93
photosynthesis, 90-91
precipitation, 91
products, 96–98
proteins, 88, 90, 94–98
soils, 92
sowing, 92–93
squalene, 95, 98
starch, 90, 95–97
storage, 93–94
temperature, 91–92, 97
transpiration, 91
vitamins, 94
yields, 90, 92–94
Groundnut, 31, 201–202
aflatoxins, 211
climates, 208
cultivars, 209
cultivation practice, 209–210
development, 208, 210
diets, 207
fertilizers, 209
genotypes, 208, 210
germplasm sources, 209
growth, 207–208, 210
LER, 209
morphology, 208
nutritional value, 210
oils, 207, 210
organic farming, 210
production, 207–210
proteins, 207, 210
soils, 208–209
sowing, 209-210
uses, 210–211
yields, 209–210

**H**

*Helianthus anuus*, 31–32
*Helianthus tuberosus*, 31, 32, 186, *see also* Jerusalem artichoke
*Helminthosporium papaveris*, 156
*Hibiscus* spp., 32
Hemp, see Industrial and edible–seed hemp
Homemade food
amaranths
popped seeds, 229
bread, 230
pasta and noodle products, 230
buckwheat
baked kasha, 227
breads, 225–226

cakes, 229
chicken filled with duck or turkey buckwheat groats, 227
cooked rolls, 228–229
dumplings, 227
Japanese soba–gaki, 228
kasha, 226
noodles, 228
pancakes, 229
salads from groats, 227
soups, 227
millets
kasha with ground meat, 230
Howard, 5
Humus, 56, 70–72, 131–132, 174, 192, 209

**I**

IFOAM, 2–4, 7, 13, 25
Indian millet, 30, 32, *see also* Koda millet
Industrial and edible–seed hemp, 30
anatomy, 166–167
botany, 166
brewages, 164
crop rotation, 172
cultivars, 166,170–175, 177
cultivation, 166–167, 172-174
delta-9-tetrahidrokanabinol (THC), 166, 177
development, 165, 167, 170–171, 174–175
ecology, 170
ecotypes, 166–167, 169–170, 172
environmental conditions, 175
fats, 169
fibers, 164–165, 167–169, 170–176
female plants, 166-167, 169, 171, 176
fertilization, 167, 169, 173
foods, 175, 177–178
functional foods, 177–178
growth, 165–167, 170, 172–174, 176
growth stages, 167, 170, 172, 176
harvest, 169, 174, 176
health, 177–178
history, 163–165
industrial processing, 166, 178
male plants, 166-167, 169, 170–171, 176
maturity, 168–170, 174–176
medicine, 165, 177
morphology, 166, 169, 176
nutrients, 168, 171–173
oils, 165–166, 172, 174–175, 177-178
organic products, 175
photoperiod, 172
plant population, 167, 168, 174

processing, 166, 174–175
production, 164-166, 168, 171–174, 176–177
prohibition, 166
proteins, 169, 177
seeds, 166, 169–177
sowing, 167–168, 171, 173–174
stems, 167–176
storage, 174, 177
textiles, 175–176
utilization, 175
weeds, 172–174
yields, 167, 169–176
Intercropping, 27, 33, *see also* Organic crop production
Intermediate wheatgrass, 31, 58
International Federation of Organic Farming Movements, *see also* IFOAM
Inspection, 14
*Ipomoea batatas*, 31, 189, *see also* Sweet potato
Italian millet, 32, *see also* Foxtail millet

**J**

Japanese barnyard, 31, 123
Jerusalem artichoke, 31, 32
    diets, 188
    environments, 187
    growth, 186–187
    harvest, 187
    industry, 189
    organic cultivation, 187
    soils, 187-188
    uses, 186–189
    yields, 186–189
Jungle rice, 32, 123, *see also* Barnyard millet

**K**

Kamut, 30, 51, *see also Triticum turgidum*
Koda (i.e. also kodo) millet 31, 109, 123

**L**

*Lactobacillus plantarum*, 75
*Lathyrus* sp. 201
Land equivalent ratio (LER), 209
Legislation, 13–14, *see also* Organic food
Legumes, *see also* Chickpea, Groundnut, Soybean and Vigna
    amino acids, 202
    climates, 202
    grain legume species, 201

insects, 201
nutrition, 202
organic farming, 2002
organic manure, 201
pathogens, 201
proteins, 202
soils, 201
*Leptosphaeria maculans*, 139
Liebig, 5
*Limnathes*, 32
*Linum usitatissimum*, 30, 32, 179, *see also* Flax
Little millet, 31, 32, 124,
*Lotus* sp., 201
Lupin, 32
*Lupinus* spp., 32

**M**

Management, 2, 4–5, 10–11, 19–21, 23–26, 28, 33
*Medicago* sp., 201
*Meligethes assimilis*, 149
Millets, 65,
    Barnyard millet, 123
    Finger millet, *see also* Finger millet
    food processing, 109
    Foxtail, *see also* Foxtail millet
    gluten-free foods, 109
    Koda millet, 123
    Little millet, 124
    nutritive potential, 109
    Pearl millet, 117–119
    Proso millet, *see also* Proso millet
    production, 109
    White fonio, see aslo White fonio
Multifunctionality, 17, se also Organic agriculture
Müller, 5

**N**

National Organic Program, 14, *see also* NOP
NOP (National Organic Program), 22

**O**

Oil (seed) poppy, see Garden poppy
Oil (seed) pumpkins, 30
    alternative medicine, 127, 135
    aromas, 136
    botany, 127
    climates, 130–131
    clinical studies, 135

cultural practice, 131–133
diseases, 130–132
fertilization, 132
fertilizers, 131
gastronomy, 127, 135
growth conditions, 131
growth stages, 127, 132
harvest, 128–129, 133–134
health, 132, 135
maturity, 128
nutritional value, 134
oils, 127, 134–136
pharmacology, 127, 135
phenological stages, 128
plant population, 132
plug trays, 133
pressure, 134, 136
production, 127–128, 130–131, 133
soils, 131, 133
sowing, 131–132
species, 127–128
storage, 134
terms, 127
transplants, 132–134
vitamin E, 127, 134
weeds, 133
yields, 129, 131-134
Oil seed rape, 30
Organic agriculture
  see Biodiversity
  data, 9–10, 20, 26
  definitions, 2, 4
  multifunctionality, 17
  rules, 6–8, 12–14, 22, 29
  sociological aspects, 17
  environments, 1–6, 8–9, 12, 15–21, 26, 28, 30, 33
Organic crop production, 22-23, 25, 30, *see also* chapters of each crop
  see rules of Organic agriculture
  organic management, 10, 23, 25–26
  see Crop rotation
  intercrops, 27–28, 33
Organic farming,
  certification, 7–8, 10–11
  legislation, 6, 13, 19–20, 22, 25
  regulations, 7, 11, 13, 24
  standards, 8, 9, 11, 55
  system, 11, 26
Organic food, 10, 12–17, 29 *see also* Homemade food
  quality, 12, 18, 30, *see also* chapter of each crop
  processing, 16, 87–88, 97, *see also* chapters of each crop

Organic products, 6–11, 13, 16, 88, 123, 149, 175, *see also* chapters of each crop
*Ornithopus* sp., 201
*Orobanche crenata*, 207
*Onobrychis* sp., 201

# P

Panicum
  sp.
  *germanicum*, 111
  *italicum*, 32, 111, 114, *see also* Foxtail millet
  *miliaceum*, 30, 32, 109, 111, *see also* Proso millet
  *sumatrense*, 31–32, 109, 124, *see also* Little millet
Papaver
  sp.
  *rhoeas*, 150
  *setigerum*, 150
  *somniferum*, 30, 32, 150, *see also* Garden poppy
*Paspalum scrobiculatum*, 31–32, 109, *see also* Koda millet
Pearl millet, 31–32
  climates, 117
  cultural practice, 117
  diseases, 117
  foods, 119
  growth, 117
  nutritional value, 118
  origins, 117
  processing, 119
  production, 117, 119
  utilization, 119
  vegetation period, 117
  yields, 117
*Pennisetum glaucum*, 31, 32, 109, 117, *see also* Pearl millet
*Pennisetum purpureum*, 93
Peronospora
  sp.
  *arborescens*, 156
  *camelinae*, 139
*Perrisia papaveris*, 156
*Phalaris arundinacea*, 51
*Phyllotreta* ssp., 149
  *cruciferae*, 139
*Pythium* spp., 149
Percentage of live seed emergence (PLSE), 91, 93
Policy
  see CAP
  rural development, 20, 21

Poppy, 30, *see also* Garden poppy
Proso millet, 30
   botany, 111
   cultivars, 111
   cultural practice, 112
   draught, 114
   environments, 111, 113
   fertilization, 113
   foods, 110, 114
   growth, 112
   maturity, 112–114
   morphology, 111
   nutrients, 112–113
   production, 110, 112, 114
   soil cultivation, 113
   transpiration coefficient, 112
   utilization, 114
   weeds, 112–114
   yields, 110–114
Production, see Organic crop production
*Pseudomonas fluorescens*, 118
*Psylliodes* spp., 149
Pumpkins, see Oil (seed) pumpkins
Puccinia
   *striiformis*, 39
   *recondita*, 39

# Q

Quality, see Organic food
Quinoa, 31
   amino acids, 78, 85–86
   botany, 79
   biodiversity, 83
   breads, 87–88
   carbohydrates, 84
   climatic conditions, 79, 84
   composition, 78, 84, 85
   crop rotation, 83
   cultivation, 78, 83
   diseases, 83, 87
   environmental conditions, 81
   emergence, 79, 81, 83
   fats, 84–85
   fibers, 84–85
   foods, 78, 79, 87–88
   genotypes, 79–81, 84
   glutens, 85, 87–88
   growth, 80–83
   harvest, 78, 81, 84
   health, 84
   homemade food, 87
   insects, 83, 87
   irrigation, 82

   maturity, 81, 83–84
   minerals, 78, 85–86
   morphology, 79
   moths, 83
   nutrients, 81–82
   nutritional value, 78, 84
   organic cultural practice, 83
   pests, 83
   plant population, 79, 84
   processing, 80, 83–84, 87
   producers, 78, 82
   production, 78–84
   proteins, 79, 84–86, 88
   saponins, 87-89
   soils, 79, 81–83
   sowing, 79, 81, 83
   taxonomy, 79
   tocopherols (vitamin E), 85
   vegetation period, 81
   yields, 78, 81, 83–84
   weeds, 79, 83

# R

*Ricinus communis*, 32
*Rhizoctonia solani*, 149
Rules, 21–24, *see also* Organic crop production
Rusch, 5

# S

Safflower, 31
   botany, 140
   cancer, 146
   cartamin, 145
   cultural practice, 141
   cultivars, 140, 142, 144–45
   cultivation, 143–144
   diseases, 142–145
   dye (i.e.color), 139–140, 145
   ecology, 141
   emergence, 143
   fertilizers, 143
   foods, 140, 145
   growth, 140–143, 145
   harvest, 143, 145
   irrigation, 142, 144
   maturity, 143
   medicine, 140, 145
   nutrient uptake, 140
   oils, 140, 143
   organic production, 142
   origins, 140
   photoperiods, 141

phenolic compounds, 146
production, 141–142
saffron, 139–140
soils, 142–143
species, 140–141
utilization, 145
weeds, 143–144
yields, 141–145
*Secale cereale*, 52
*Septoriaa nodorum*, 39
*Setaria italica*, 30, 114, *see also* Foxtail millet
*Simmondsia chinensis*, 32
*Sinapis alba*, 30, 146, *see also* White mustard
Sociological aspects, 17–19, *see also* Organic
    agriculture
*Sorghum* spp., 32
Soybean,
  amino acids, 213
  climates, 213, 215
  cultivars, 213
  cultivation, 215–216
  fertilization, 216
  genotypes, 213
  growth, 215
  harvest, 217
  maturity, 215, 217
  morphology, 213
  nitrogen bacteria, 213, 216
  organic production, 211, 216
  plant diseases, 217
  precrops, 215
  production, 214, 217
  products, 211–213
  proteins, 212
  soils, 216
  sowing date, 216
  storage, 217
  temperatures, 214
  transpiration coefficient, 215
  weeds, 216–217
Spelt, 30
  amino acids, 43–44
  botany, 38
  breads, 38, 45–47
  carbohydrates, 43
  chemical composition, 42
  coffe substitute, 45
  cultivars, 39, 42-43, 45–46
  diseases, 38–39, 41, 45
  emergence, 40, 42
  fatty acids, 42, 45
  fats, 42–43, 45
  fertilization, 40
  fibers, 42–43
  flours, 43, 45

genotype, 38, 42, 46
gliadins, 45
gluten-free food, 45
gluten-index, 46
green spelt seeds, 45
grind, 46–47
growth, 38–39, 41–42
grünkern, 45–46
harvest, 38–39
health, 42
huskers, 45
interrow spacing, 41-42
maturity, 45, 47
morphology, 39
nutritional value, 42–44
nutrients, 40
organic cultivation practice, 39
organic flakes, 47
organic diets, 47
origin, 37–38
processing, 45–46
production, 38–39, 41, 46, 48
proteins, 42-43, 46
seeding (i.e. sowing) rates, 41
sowing date, 40,
storage, 42
threshers, 45
tillage, 40–41
tiller, 39, 41
varieties, 38, 39, 43, 45
vitamins, 43, 45
winter, 40–43
soups, 45, 47
summer, 40, 43, 50
systematic, 38
Standards, 8–9, *see also* Organic farming
Steiner, 5
Subsidies, 6, 19
Sunflower, 31
Sustainable production, 1, 4, 12, 18, 25
Sweet potato, 31
  antioxidative activity, 194
  botany, 190
  climates, 189, 191, 193
  cultivars, 190, 192–193
  cultivation, 191–192
  development, 191–192, 194
  food processing, 193
  growth, 191–192
  health value, 194
  industry, 193
  maturity, 192
  mycorrhiza, 192
  nutrition, 192–193
  production, 189, 190, 193

reproduction, 190, 193
ripeness, 192
soils, 190–193
storage, 192–193
tubers, 190–193

# T

*Tef*, 93
THC see Delta-9-tetrahidrokanabinol
*Thinopyrum intermedium*, 58, *see also*
          Intermediate wheatgrass
*Trifolium* sp., 201
*Trigonella*sp., 31, 201
*Triticale see also* Triticale
   *aestivum*, 53
   *durum*, 53
   *trispecies*, 53
Triticale, 31
   amino acids, 57
   bread composition, 52, 57–58
   countries, 54
   crop rotation, 55, 57
   cultivars, 52–58
   dry conditions, 54
   environments, 54–55, 57
   fertilization, 57
   fertilizers, 59
   flour, 57–58
   geographical position, 55
   germination, 55–56
   gluten, 57–58
   grain deformation, 54–55
   grain quality, 54
   grain weight, 55
   growth, 54–56
   growing conditions, 56
   harvest index, 56–57
   health, 57
   history, 53
   maturity, 54, 57
   morphology, 54
   nitrogen, 57
   nutrition, 57
   nutrients, 57
   organic cultivation, 55
   products, 53, 57
   proteins, 58
   quality, 54, 57–58
   sea level, 56
   selection, 53, 57
   senescence, 55
   spring cereal, 54
   soils, 56

   sowing, 55–56
   processing, 57
   production, 52–57
   taxonomy, 53
   temperatures, 52–56
   uses, 53, 55–58
   water supply, 55
   winter, 54
   weeds, 56–57
   yields, 52–57
*Triticosecale*, 31, 52, *see also* Triticale,
Triticum
   sp.
   *aegilopoides*, 47
   *aestivum*, 38, 53–54
   *arduini*, 38
   *beoticum*, 47
   *dicoccum*, 31, 37–38, 50, see aslo Emmer
   *durum*, 37, 54
   *monococcum*, 30, 37–38, 47, *see also* Einkorn
   *sinskajae*, 38
   *spelta*, 30, 37–38, *see also* Spelt
   *tauschii*, 38
   *thaoudar*, 47
   *turgidum*, 31, 38, 51, *see also* KAMUT[R]
   *urartu*, 47

# U

Utilization, 28–32, *see also* Alternative crops

# V

*Vicia* sp., 201
Vigna, 217–220
   azuki beans, 218–220
   climates, 217–218
   growth, 217, 220
   morphology, 218–220
   mungo beans, 218–219
   nitrogen, 220
   producers, 217
   production, 217–218, 220
   proteins, 217
   species, 218-220
   weeds, 220
   yields, 218
Vigna
   sp., 32, 201
   *aconitifolia*, 218
   *angularis*, 217–218
   *cylindrica*, 220
   *glabrescens*, 219
   *mungo*, 218

*radiata,* 217–219
*subterranea,* 220
*umbellatea,* 218
*unguiculata,* 218

# W

Wheats
  hulled, 38
White fonio
  biology, 121
  carbohydrates, 122
  cultivars, 122
  cultivation, 122
  foods, 120–122
  mineral content, 122
  processing, 121, 123
  products, 123
  production, 121
  proteins, 122
  soils, 122
  utilization, 122
  yields, 122
White mustard, 30
  aromas, 146
  botany, 147
  biodiversity, 149
  cancer, 147
  climates, 148
  cultivars, 146-148
  diuretics, 146
  genotypes, 147
  growth, 147-148
  grinding seeds, 149
  harvest, 149
  honey yield, 147
  lubricants, 146
  maturity, 149

medicine, 146
melliferous plants, 147
oils, 146
organic products, 149
origins, 146
pathogens, 149
photoperiodism, 149
processing, 149-150
production, 148, 150
resistance to beet nematodes, 148
salads, 150
soils, 148
utilization, 146
Wild rice, 32
  biology, 98
  cultivation, 99
  diseases, 100
  flooded fields, 99
  food, 100
  growth , 99
  harvest, 99
  maturity, 99
  nutritional, 99
  plant population, 99
  production, 98
  proteins, 99
  utilization, 99
  yields, 100

# Z

Zizania
  sp.
  *aquatiaca,* 98–99
  *palustris,* 31, 98–99, see Wild rice
  *latifolia,* 98
  *texana,* 98